Cambridge El

Elements in the Philoso
edited by
Keith Frankish
The University of Sheffield

PHILOSOPHY OF NEUROSCIENCE

William Bechtel
University of California, San Diego
Linus Ta-Lun Huang
University of Hong Kong

CAMBRIDGE
UNIVERSITY PRESS

University Printing House, Cambridge CB2 8BS, United Kingdom

One Liberty Plaza, 20th Floor, New York, NY 10006, USA

477 Williamstown Road, Port Melbourne, VIC 3207, Australia

314–321, 3rd Floor, Plot 3, Splendor Forum, Jasola District Centre, New Delhi – 110025, India

103 Penang Road, #05–06/07, Visioncrest Commercial, Singapore 238467

Cambridge University Press is part of the University of Cambridge.

It furthers the University's mission by disseminating knowledge in the pursuit of education, learning, and research at the highest international levels of excellence.

www.cambridge.org
Information on this title: www.cambridge.org/9781108931502
DOI: 10.1017/9781108946964

First published 2022

A catalogue record for this publication is available from the British Library.

ISBN 978-1-108-93150-2 Paperback
ISSN 2633-9080 (online)
ISSN 2633-9072 (print)

Philosophy of Neuroscience

Elements in the Philosophy of Mind

DOI: 10.1017/9781108946964
First published online: February 2022

William Bechtel
University of California, San Diego

Linus Ta-Lun Huang
University of Hong Kong

Author for correspondence: William Bechtel, bill@mechanism.ucsd.edu

Abstract: This Element provides a comprehensive introduction to philosophy of neuroscience. It covers topics such as how neuroscientists procure knowledge, including not just research techniques but the use of various model organisms. It presents examples of knowledge acquired in neuroscience that are then employed to discuss more philosophical topics such as the nature of these explanations, the different conception of levels employed, and the invocation of representations in neuroscience explanations. The text emphasizes the importance of brain processes beyond those in the neocortex and then explores what makes processing in the neocortex different. It considers the view that the nervous system consists of control mechanisms and considers arguments for hierarchical versus heterarchical organization of these control mechanisms. It concludes by considering the implications of findings in neuroscience on how humans conceive of themselves and practices such as embracing norms.

Keywords: neuroscience, explanation, methods of inquiry, levels, representations

ISBNs: 9781108931502 (PB), 9781108946964 (OC)
ISSNs: 2633-9080 (online), 2633-9072 (print)

Contents

1 Introduction: What Is Philosophy of Neuroscience?

Neuroscience is an interdisciplinary scientific inquiry of neural processes. We commonly identify these processes with the brain, but in fact neurons are distributed throughout animal bodies (we have over 500 million neurons in our guts, constituting what is referred to as the *enteric nervous system*). The reason neuroscience is interdisciplinary is that the research techniques of different disciplines are required to understand neural processes. Most obvious are anatomy and physiology, which address the structure and operation of neurons and the larger structures built out of them. Genetics has proven extremely important both for characterizing neural processes but also for altering them (alteration is required in any experimental study). Often, given the complexity of neural processes, it is helpful to model them computationally, giving computer science an important role in the interdisciplinary mix. One of the reasons neural processes have drawn so much interest is their role in behavior and cognitive function; accordingly, psychology and cognitive science are also important contributors to neuroscience.

What motivates philosophers to examine neuroscience? There are a variety of motives. One is the thought that knowing about the brain tells us important things about ourselves that are relevant to other philosophical inquiries about topics such as whether human action is free, whether we can know our world, and what it is to be conscious. This pursuit often goes by the name *neurophilosophy*, a term introduced by P. S. Churchland (1986). We will take up some questions posed by neurophilosophy in the last section. A second motive is to apply philosophical methods to problems in neuroscience. Philosophers often have skills that enable them to generate hypotheses, integrate different findings, and clarify concepts in ways that are useful to neuroscience. This pursuit might best be termed *philosophy in neuroscience* (the distinction between philosophy of and philosophy in was developed by Brook in the context of cognitive science; see Brook, 2009). This is covered in Sections 9 and 10. Our primary focus will be on a third approach that investigates how neuroscience functions as a science: Which methods are employed? Which organisms are studied? What does a neuroscientific explanation look like? Since these are philosophy of science questions about neuroscience, this approach is best labeled *philosophy of neuroscience.*

Since neuroscience constitutes the subject matter of our inquiry, we will at various points present some of the knowledge developed in neuroscience. Thus, in Section 2, we introduce neurons and neural processes, and in Section 5, we offer four vignettes illustrating what neuroscience has learned about mental processes: situating ourselves in time, navigating through space, seeing the world, and making decisions. We will review this content in later sections. In Sections 3 and 4, we address fundamental issues about how neuroscientists gain knowledge: how they study neural processes, and whose neural processes they study. A major aim of neuroscientists is to offer explanations for behavior and cognition, and Section 6 will offer accounts of what is required of an explanation. Sections 7 and 8 focus on more specialized issues of neuroscience explanations: the levels at which explanations are offered and whether explanations should attribute representations to neural processes.

Both in neuroscience and in philosophy it has been common to adopt a cortico-centric view of the brain, but in fact there is extensive research in neuroscience on subcortical areas. Subcortical processing is extremely important in determining how we behave. This is significant since cortical areas constitute a different type of neural processing system than subcortical areas, and in Section 9, we focus on what is distinctive about the neocortex in particular. We then turn to the question of how the whole brain is organized. It is often viewed as organized as a hierarchy with the neocortex at the top, and indeed, one part of the neocortex, the prefrontal region, at the very top, operating as a central executive. In Section 10, we contrast this with a heterarchical perspective that views neural processes as organized in an interactive network, with different regions exercising control over different aspects of behavior and cognition. Finally, in Section 11, we pull from various topics addressed in earlier sections to address the neurophilosophical question of what neuroscience has to teach us about ourselves as agents in the world.

2 What Are Neurons and Neural Processes?

Most people have seen multiple (typically idealized) pictures of the human brain as it would appear if one opened up the skull. The first thing one notices is a highly convoluted gray structure (at the top of Figure 1) in which the projecting areas are known as gyri and the indented areas as sulci. This structure, known as the *neocortex*, is often divided into four lobes: frontal, occipital, parietal, and temporal. As the part of the brain that has most expanded in the lineage of primates, including us, it has assumed a central focus in much philosophical theorizing. However, as the characterization of it as *neo* suggests, there is more to the cortex (often termed the *cerebral cortex*), including very important structures such as the hippocampus. The term *cortex* is derived from the Latin term for the bark of a tree and, as that suggests, it refers just to the outer structure. There is much of the brain beneath the cortex.

In this Element, we seek to avoid the all too frequent cortico-centric take on the brain by focusing as much on what is beneath the cortex and the

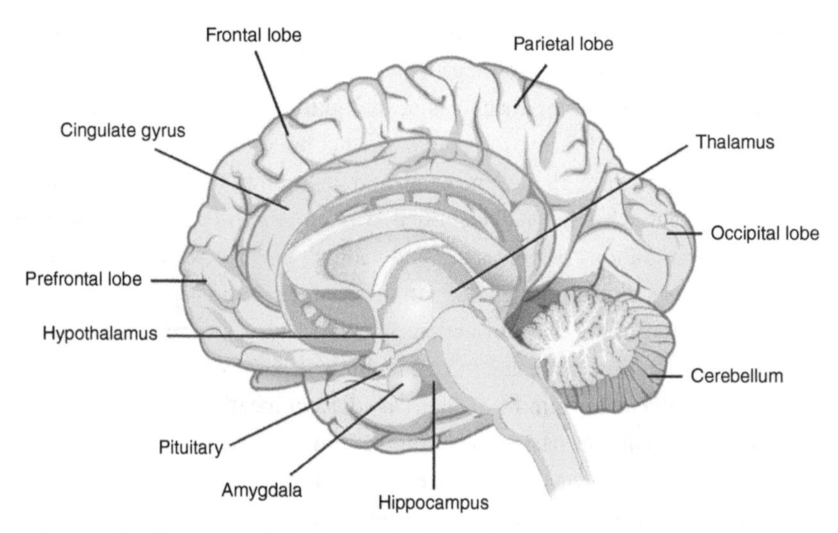

Figure 1 Major areas of the human brain. Adapted from OpenStax College – Anatomy & Physiology, Connexions website: http://cnx.org/content/col11496/1.6/, June 19, 2013., distributed under CC BY 3.0, https://commons.wikimedia.org/w/index.php?curid=30148029.

philosophical questions that those areas engender. By taking into account subcortical brain regions and their role in behavior and cognition, we will be in position, by Section 9, to address what is different about the neocortex and how it provides humans with distinctive cognitive abilities. Even as it enables these distinctive abilities, the neocortex does so through interacting with subcortical regions. For now, we start from the building block of all nervous tissue, the neuron, and then consider ways in which neurons are organized.

2.1 The Neuron

The neuron is a specialized type of cell. Although neurons are too small to be seen with the naked eye, ancient anatomists did observe nerves (bundles of neurons), recognized their importance in transmitting signals through the body, and speculated on their constitution. Most hypotheses viewed them as functioning much like blood vessels, with very fine matter (animal spirits, where *spirit* refers to fine matter, as in *spirits of alcohol*) flowing through them. Only in recent centuries did researchers ascertain that neurons transmit electrical current.

The research that would reveal electrical transmission of neurons began around 1600, when investigators (and the lay public!) began experimenting with electrical shocks, including those generated by friction machines. Many were fascinated by how these could cause muscle contractions. Based on extensive experiments with frog legs, involving the elicitation of muscle contraction by spark-generating machines or by lightning, Galvani (1791) argued that muscles possessed their own source of what he termed *animal electricity*. Extensive research through the nineteenth and first half of the twentieth centuries revealed that what he had identified was an electrical potential due to different concentrations of potassium and sodium ions across the membranes of both muscles and neurons. Changes in these concentrations propagate along the neuron, often in the form of action potentials, also referred to as *spikes*. Action potentials are large changes in the membrane potential at one location on the membrane that cause similar changes at adjacent locations, creating a wave of electrical current that passes along the neuron until it reaches the synaptic terminal at the end of the neuron. Figure 2 shows the now canonical representation of the action potential, which begins with the neuron negatively polarized (approximately −70 mV; referred to as the *resting potential*). When a stimulus is sufficient to push the potential above threshold, it rapidly and temporarily depolarizes to approximately +40 mV before repolarizing. When neurons propagate action potentials, they are often said to *fire*, capturing the fact that action potentials represent relatively discrete signals propelled along neurons.[1]

During the same period (the nineteenth century), other researchers were examining biological tissues with the light microscope. They identified what are termed *cells* and advanced the theoretical framework in which cells are the basic living units. Adding stains enabled researchers to see the projections – axons and dendrites – that differentiate neurons from other cells. One stain, a silver nitrate stain introduced by Comillo Golgi, was particularly informative since, for reasons still not understood, it only stains some neurons in a preparation. This makes it possible to visualize individual neurons. Golgi, however, did not interpret what he saw as individual cells but rather as a continuous reticular network of nerve tissue. Adopting Golgi's stain and visualizing such things as developing or degenerating nerve fibers, Santiago Ramón y Cajal concluded that the network was not continuous; rather, there were gaps between projections from different nerve cells. Drawing upon Cajal's work, Waldeyer invented the term *neurone*, now *neuron*, and articulated the *neuron doctrine* according to which discrete neurons are the units of nerve tissue. At the end of the nineteenth century, the dispute between Golgi and Cajal was very contentious, and even as both were awarded the Nobel Prize in 1906, Golgi

[1] Not all neurons generate action potentials. Some transmit graded potentials. Instead of discrete, digital signals, they generate responses of varying magnitude. An important advantage of signaling with action potentials is that they can be maintained over long distances without loss of content.

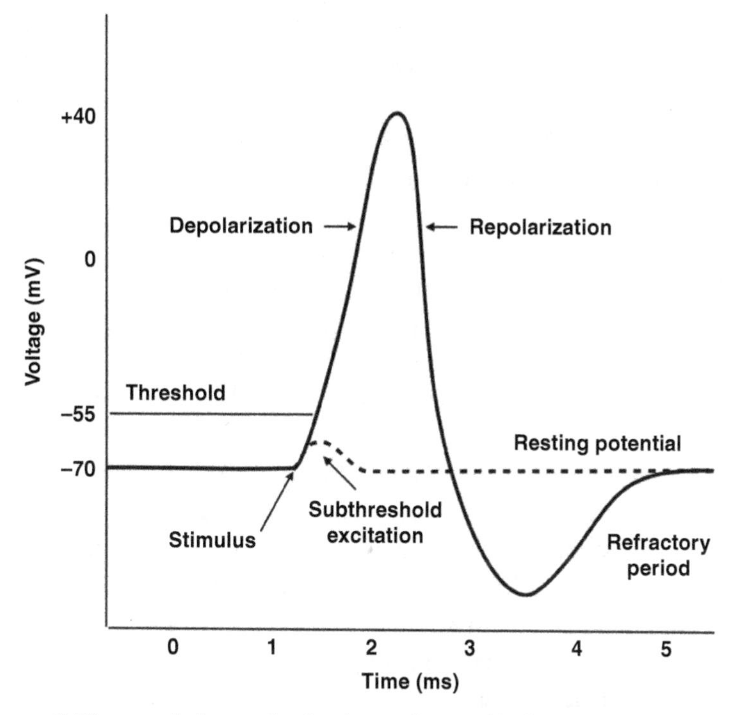

Figure 2 Characteristic graph of voltage changes during an action potential. Adapted from en:User:Chris 73, updated by en:User:Diberri, converted to SVG by tiZom – distributed under CC BY-SA 3.0, https://commons.wikimedia.org/w/index.php?curid=2241513.

continued to defend the reticular view, arguing that only if nerves consisted of an interconnected network would they be able to communicate messages through the body. The conflict between Golgi and Cajal is an illuminating example of how skilled observers can reach conflicting conclusions and how such conflicts are resolved (for discussion and details, see Mundale, 2001; Shepherd, 2016).[2]

Cajal argued that the two types of processes extending from the neuron cell body play different roles. He interpreted the typically short and highly branching structures, known as *dendrites*, as receiving inputs from other neurons, and the longer, less branched structures, known as *axons*, as carrying output to other cells. He supported this by the observation that sensory neurons have their dendrites oriented toward the sense organ (e.g., the eye) and axons oriented

[2] One might think such a schism could be resolved simply by looking carefully through the microscope, but the gap between neurons is too small to be seen with the light microscope. When the electron microscope was applied to nerve tissue in the 1950s, it did reveal the gap, but ironically it also revealed the presence in some cases of direct contacts between nerve cells, known as gap junctions.

toward the brain. In making this distinction, he proposed that there was one-way transmission through the nervous system. In 1897, Charles Scott Sherrington characterized the gap between neurons as *synapses* (derived from the Greek for "to clasp"). Figure 3 shows a prototypical neuron. Although we will not develop the point, one should note that neurons exhibit enormous variety both in appearance and function.

The discovery of synapses presented a new challenge: How do signals cross the gaps between neurons? The initial assumption of many researchers was that electrical charges could jump synapses, much as sparks from a spark generator can jump to a grounded surface. A long lineage of research, especially in the first half of the twentieth century, ultimately revealed this was incorrect and transmission between neurons is chemical.

Most of the initial work that led to this conclusion focused on the junction between nerve and muscle. Around the turn of the century, a few pharmacologists and chemists began investigating substances (such as an extract from the adrenal gland initially referred to as noradrenaline and later as norepinephrine) that elicited or inhibited responses of muscles. A notable finding was the accumulation of another chemical, acetylcholine, in heart tissue when stimulated by the vagus nerve (which projects from the central brain to the heart, lung, and intestines). Many investigators, however, initially resisted the idea that acetylcholine was released by the nerve and caused contraction of heart muscles

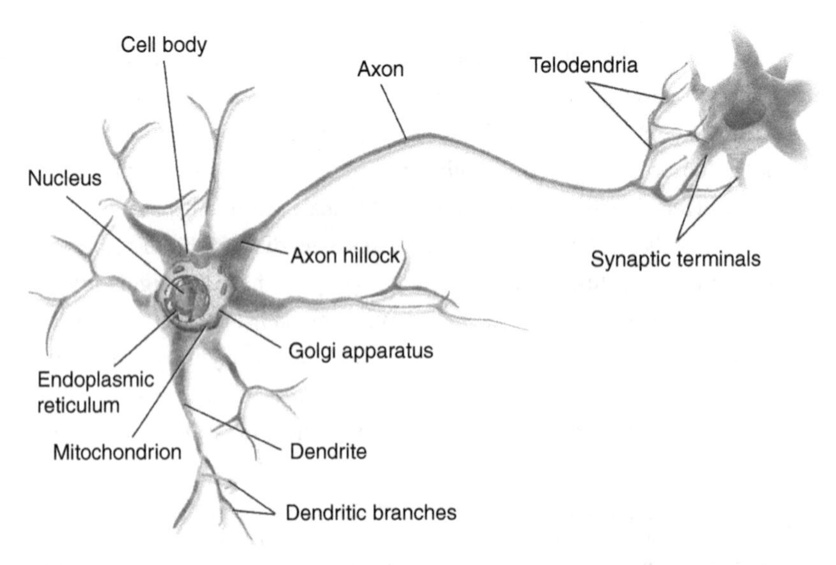

Figure 3 A prototypical neuron. Figure by Bruce Blaus, distributed under CC BY 3.0, https://commons.wikimedia.org/w/index.php?curid=28761830.

to slow down. In 1920, Otto Loewi provided compelling evidence that acetylcholine was released by nerves by bathing the heart of one frog in liquid and stimulating the vagus nerve. Once its heart contractions slowed, he transferred the liquid to the heart of another frog whose vagus nerve had been removed. Heart contractions in that frog also slowed.

Although this provided compelling evidence that chemicals released by neurons act on internal muscles such as those in the heart, many resisted the idea that chemicals transmit signals between two neurons or between neurons and skeletal muscles. Chemical signaling, it was thought, was too slow. This ensuing conflict came to be known as the war of the soups (advocates of chemical transmission) and the sparks (advocates of direct electrical transmission). (For an engaging analysis of the conflict, see Valenstein, 2005.) In the wake of the victory by the soups, hundreds of chemicals, referred to as neurotransmitters, have been discovered and neuroscientists have developed an understanding of how they are synthesized and released from one neuron and, by binding to receptors, generate changes, including action potentials, in other neurons.

In most cases, neurotransmitters bind to a receptor in the postsynaptic cell and serve either to depolarize it (thereby increasing the likelihood that it will generate an action potential) or further polarize it (thereby inhibiting it). Any excess is typically quickly broken down and the components recycled. Some neurotransmitters, referred to as *volume transmitters* or *neuromodulators*, disperse widely and serve to modulate the behavior of neurons that have the appropriate receptors. We noted that neurons come in a huge variety. An important type of variation involves the neurotransmitters that they release or to which they respond.

2.2 Foundational Neural Structures: Nerve Networks

As important as individual neurons are, they typically carry out their activities as parts of collectives. Hence, in this and the following sections, we focus on some important ways in which neurons combine into larger structures.

Given that dendrites receive signals and axons send out signals, it is plausible to view neurons as having evolved to connect sensory and motor processes. Philosophers Keijzer, van Duijn, and Lyon (2013) have challenged that view, arguing instead that the first function of neurons was to coordinate muscles. They refer to their proposal as the *skin–brain hypothesis* and appeal to jellyfish to illustrate it. Jellyfish belong to the phylum *Cnidarian*, which differentiated from other animals between 500 and 700 million years ago and is thought to be representative of early evolved animals. A prominent feature of jellyfish is the

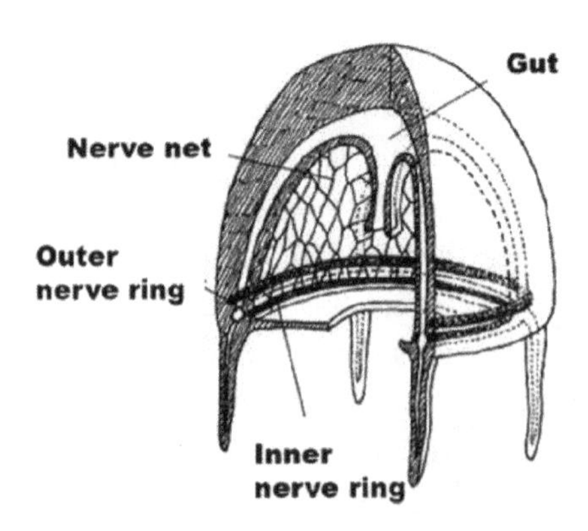

Figure 4 The nerve net and nerve rings in a typical jellyfish. Reprinted from Koizumi (2016) by permission of Springer Nature.

bell (Figure 4). By rhythmically contracting the bell, jellyfish are able to swim upward, after which they drift downward, collecting food in their gut that is positioned inside the bell (a process aided by tentacles that project from the bell). Contractions are produced by two layers of contractile epithelial cells (proto muscles) that line the bell. To generate a rhythm, the contraction of these cells must be coordinated. To some degree, this is accomplished through gap junctions (conduits that allow small molecules to pass between cells directly) between epithelial cells. But to coordinate contractions over longer distances, jellyfish rely on a network of neurons (a nerve net) situated between the two epithelial layers whose processes reach across the bell. Whenever these processes cross each other, they create a distinctive type of synapse in which either neuron can release peptides (short chains of amino acids linked by peptide bonds) that act on ion channels on the other neuron, thereby altering its electrical activity. As a connected system, the nerve net can coordinate the activity of contractile cells. Keijzer, van Duijn, and Lyon adopt the term *skin brain* for this nerve net.

An important requirement for any organism is that it can alter its behavior to accommodate local conditions. In jellyfish this is accomplished by other neurons, including those forming the two nerve rings that encircle the bell and send signals to neurons in the nerve net. The function of these nerve rings has been extensively investigated in one jellyfish species, *Aglantha digitale*. Neurons in the inner ring function as a pacemaker for the rhythmic activity in the nerve net. The default pace of the inner ring is a slow rhythmic firing (Mackie, 2004; Satterlie, 2018). When sensory neurons in the tentacles are activated, they act on inner ring neurons, causing them to produce a faster, higher amplitude rhythm that produces faster swimming that enables to jellyfish to escape (Arkett, Mackie, & Meech, 1988). The signal from the tentacles is just one of several signals that cause the pacemaker neurons to change activity. For example, when the jellyfish is transferring food to its mouth (a process adversely affected by its usual pulsating movement), a signal is sent to stop temporarily the slow rhythms (Mackie, Meech, & Spencer, 2012).

Whether or not Keijzer, van Duijn, and Lyon's skin–brain hypothesis that neurons evolved to coordinate the contraction of muscles is correct, it captures a common feature of neural organization – the coordination of muscles through a network of neurons. Many internal organs, such as the heart, lungs, and the intestines, engage in rhythmic muscle contractions regulated by networks of neurons. The same need for coordination is manifest in skeletal muscles that move external limbs. In this case, small networks of neurons, known as *central pattern generators* (CPGs), generate patterns that orchestrate the contraction of muscles. (These networks are perhaps better characterized as *local pattern generators* because of their location in the spinal cord.) As in the jellyfish, these networks, whether acting on internal muscles or skeletal muscles, are responsive to activity in other neurons that impinge on them.

2.3 Coordinating Centers: Ganglia and Nuclei

The neurons that act on nerve networks are often organized into anatomically differentiable collectives with their cell bodies near each other and receiving inputs (either via chemical synapses or gap junctions) from each other. In invertebrates, these are called *ganglia*. Ganglia are distributed both throughout the organism and, when the organism has one, the central brain. In vertebrates, the term *ganglion* is generally reserved for structures in the periphery and the term *nucleus* is used for those in the central brain. Arendt, Tosches, and Marlow (2016) hypothesize that an organization of nerve nets and ganglia originated even before jellyfish in an ancestor of all extant animals. This hypothesized ancestor is assumed to consist of a sac with an inner and outer layer of cells, much like that found in the gastrula stage of embryonic development in contemporary mammals. They propose that the nerve net wrapped the whole body and that neurons in two regions of the nerve net were organized into ganglia serving more specialized functions. Those around the digestive opening acted to control feeding activities, while those at the opposite end, the apical pole, specialized in detecting environmental conditions. Over evolutionary time, these ganglia progressively divided into more specialized ganglia, with those at the apical pole forming the apical nervous system (ANS), which senses conditions within the organism and its environment (including light and contact with other objects) and directs actions such as feeding and locomoting. This system also determines the timing of reproductive activities. Particularly notable is that these neurons signal using volume transmitters that diffuse through the organism: serotonin to signal satiation, neuropeptide Y to signal hunger, and dopamine to indicate the presence of food locally (Voigt & Fink, 2015; Hills, 2006). The initial ganglion around the digestive opening also expanded into the blastoporal nervous system (BNS), which provides more specific control over individual sets of muscles.

Over evolutionary time, both ganglia moved to the front in bilateral organisms, creating a chimeric central brain (Tosches & Arendt, 2013) – that is, a structure composed of parts with different origins. Tosches and Arendt maintain that the dual origin of the brain is still manifest in vertebrates, including us. The ANS developed into the anterior region of the hypothalamus, a collection of nuclei that monitor the overall state of the organism and its environment, and activate activities such as feeding and reproduction, generally through the release of hormones and volume transmitters. The BNS developed into much of the rest of the central brain, including brainstem and midbrain motor control centers, the basal ganglia, and cortical areas.

Drawing upon the distinction between the ANS and the BNS, Cisek (2019) offers a compelling picture of how these neural systems work together in generating action through loops that procure information and initiate actions required for the animal to maintain itself. A sensing system that registers a condition outside of the acceptable range generates an impetus for action until the sensing system registers that the condition is again acceptable (Figure 5(a)). Figure 5(b) elaborates on this scheme in the case of feeding activity. Reduction in serotonin (5HT) levels in the ANS registers nutritional shortfall, triggering release of neuropeptide Y. If other sensory neurons indicate the presence of food nearby, they release dopamine and the two transmitters together initiate activity in BNS neurons that direct exploitation of the local environment and inhibit motor neurons that would cause the animal to move further afield. If no food is sensed, dopamine is not released, resulting in BNS neurons that direct explorative movement becoming active, and the animal begins foraging. When food is detected, dopamine is again released, resulting in exploitation of the food source. When the animal's need for nutrients is satisfied, serotonin levels increase and the animal is free to pursue other activities.

The medicinal leech provides an illustrative example of how an animal can exercise complex control over behavior through a collection of ganglia. This example also illustrates how organisms build upon the schema put forward by Cisek to achieve effective regulation of behavior. Each of the 21 segments along the leech's body contains a ganglion of approximately 400 neurons that

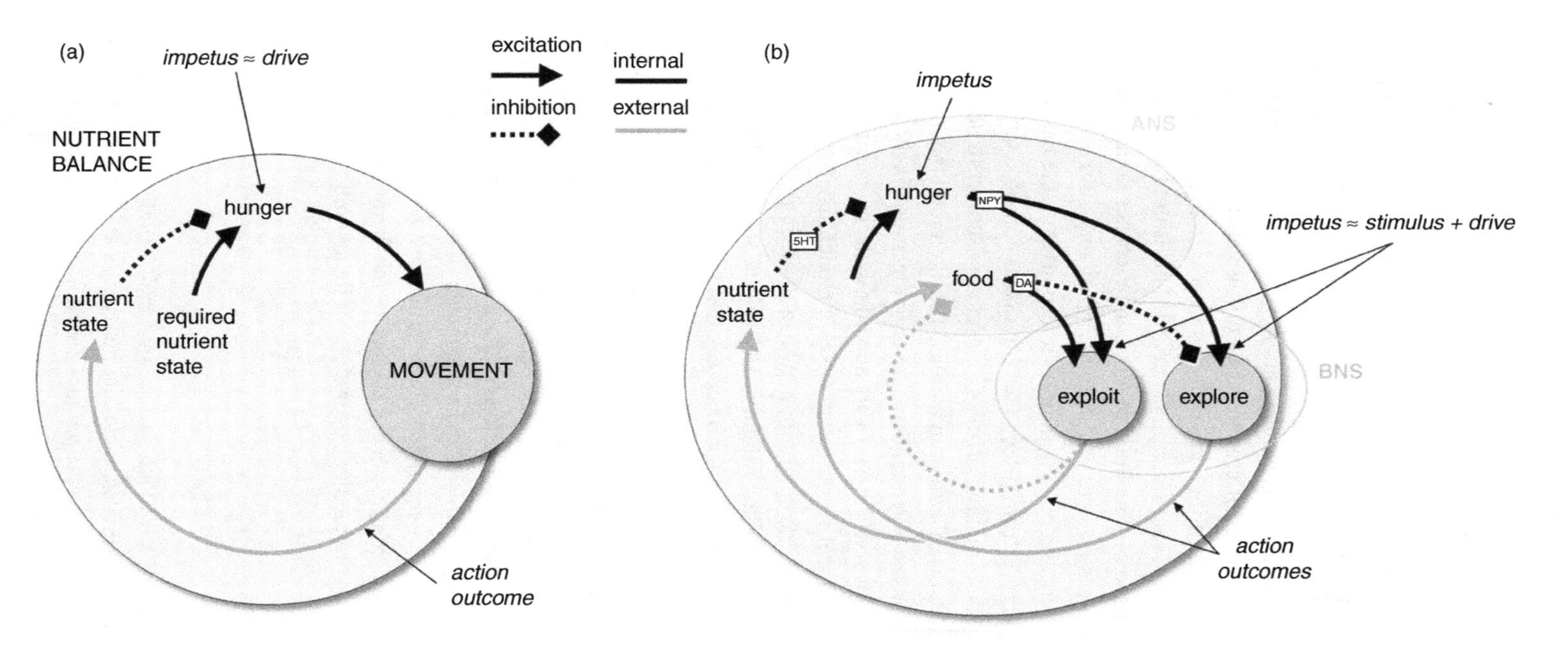

Figure 5 Collaboration of ANS and BNS in controlling feeding behavior. Taken from Cisek (2019).

makes its own decision whether to swim and crawl. A common model for collective decision-making, implemented in many human social organizations, is *winner-take-all*: each individual (in this case, each neuron) votes and the decision with the most votes determines the behavior. Briggman, Abarbanel, and Kristan (2005) demonstrate that this is not what happens in the leech. Instead, a dynamical process ensues in which neurons that are the first to become active cause the whole ganglion to settle into either an attractor for swimming or one for crawling (we discuss attractors in dynamical systems in Section 6.2). Since it won't work for one segment to try to crawl and another to swim, coordination is needed. This is not brought about by a central decision-maker in the brain (even though leeches do have a head ganglion with several sensory neurons, they don't play this role) but by each ganglion signaling others and biasing their own responses on the basis of the responses that they receive from others (see the discussion of heterarchy in Section 10). This decision-making process in leeches is subject to neuromodulation by the ANS. Baths of serotonin increase the likelihood that the leech will swim; dopamine, on the other hand, makes crawling more likely (Crisp & Mesce, 2006; Puhl & Mesce, 2008). (This accords with the role played by dopamine in exploiting local resources in Cisek's model discussed earlier.) Gaudry et al. (2010) found that serotonin may also figure in the modification of behavior during feeding. When sanguivorous (blood-sucking) leeches are feeding, serotonin blocks the sensory receptors that trigger both swimming and crawling.

In both higher invertebrates and in vertebrates, collections of neurons in the central brain exercise control over activities that are regulated by peripheral ganglia in the leech. There are two caveats, however, that should be kept in mind – ganglia and pattern generators in the periphery still perform important roles in determining the character of actions that are executed. As Sterling and Laughlin (2015) develop, organisms take advantage of local processing as much as possible. Even the brain in which processes are centralized is a collection of multiple ganglia/nuclei that each carry out their own processing in semiindependence and couple their operations in the overall control of the organism.

2.4 Specialized Information Processing Areas: Laminar Structures in the Cortex

While all brain areas have expanded in the course of evolution, the greatest change in the evolution of primates, including humans, is the massive increase from the small structure, known as the *pallium*, found in early vertebrates to the large cortex (including not just the neocortex but also the hippocampus) that dominates our brains. Cortical areas exhibit a different mode of organization than the ganglia/nuclei in the rest of the brain, a laminar structure in which neurons are organized into layers. The neocortex, in particular, is organized into six layers. At the beginning of the twentieth century, Brodmann (1909/1994) described these layers in slices of cortex prepared with the same stains used to identify neurons. Different stains resulted in different appearances but each of them revealed six layers (Figure 6(a)). Brodmann showed that different layers consist of types of neurons that are distinguished in terms of size, patterns of axons and dendrites, and so on. As is suggested in Figure 6(a), many of the projections from individual neurons project to neurons in the layers above or below them, creating what are known as cortical columns.

One feature that stood out to Brodmann was that the layers in different parts of the neocortex are of different thicknesses and often exhibit sharp boundaries where the thickness of layers changes (Figure 6(b)). He viewed these as demarcating distinct areas of the neocortex. He numbered these in the order in which he investigated them, producing the map in Figure 7. Subsequent neuroanatomists used other measures, such as patterns of connectivity between neurons, leading to somewhat different maps. Brodmann's numbering scheme, though, remains widely used.

In Section 9, we will examine the type of processing facilitated by the arrangement of neurons in the neocortex. For now, though, we will emphasize a feature of neocortex that is often overlooked. It is common to treat the neocortex as an autonomous information-processing structure. Sensory information first arrives at primary sensory areas – for example, BA (Brodmann area) 17 for vision, BAs 3, 2, and 1 for somatosensory information. It is then processed through a variety of intermediate areas, sometimes referred to as *association areas*. Finally, motor commands are developed in the premotor cortex (BA 6), and further articulated in the motor cortex (BA 4). However, as we discuss further in Section 9, each of these areas is as densely interconnected with subcortical nuclei, especially those in the thalamus and the basal ganglia, as they are with other cortical areas, rendering the neocortex a component in a functionally integrated neural system.

2.5 Summary

In this section, we have introduced the basic components of the nervous system – neurons, nerve nets, ganglia/nuclei, and cortical sheets. Each of these will figure in our discussions. We turn first to the question of how neuroscientists acquire knowledge about these entities.

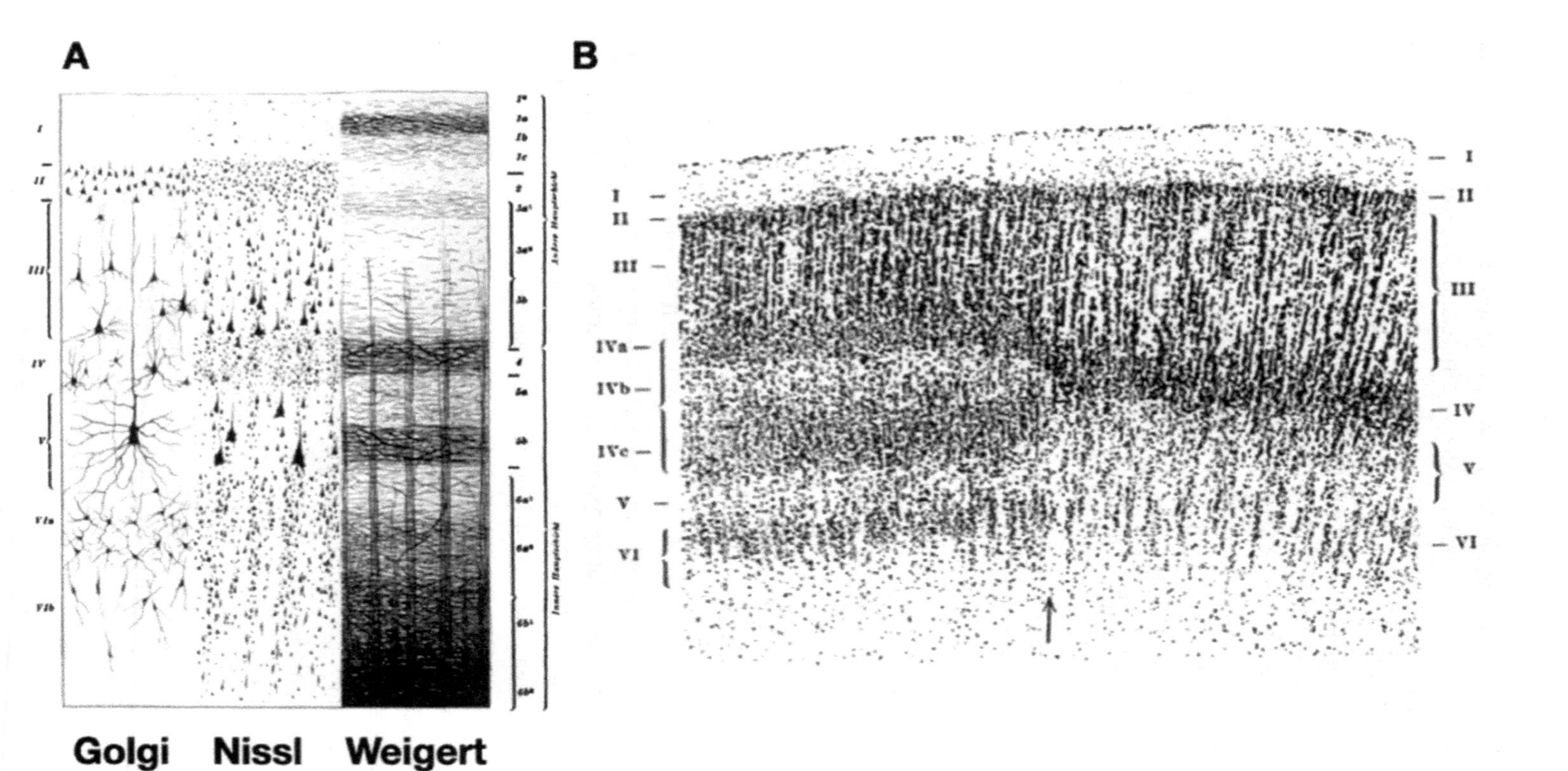

Figure 6 (a) Examples of the appearance of a typical section of neocortex with three different stains, revealing the existence of six different layers. (b) Brodmann's identification of locations (marked by an arrow) where thickness of layers changes, marking a boundary between different regions. Taken from Brodmann (1909/1994).

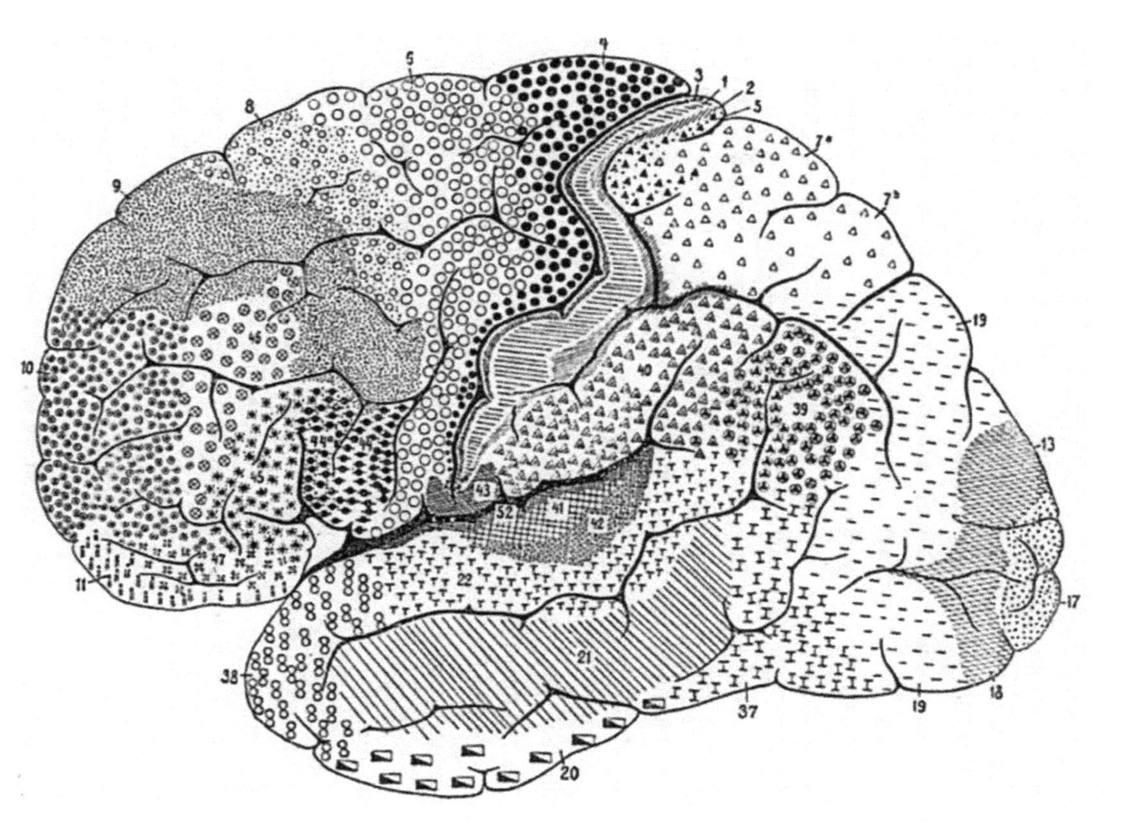

Figure 7 Brodmann's map of different areas (now referred to as Brodmann areas) in the human brain. Taken from Brodmann (1909/1994).

3 How Do Neuroscientists Learn about the Nervous System?

An important philosophical question about any field of knowledge is how its practitioners acquire and justify their knowledge claims. Although some have hoped that we could prove our knowledge claims, proof is only possible in fields such as mathematics and logic, not science. As the history of science has shown, even hypotheses that were supported by very strong evidence might later be revealed to be false. This, however, should not generate despair or the conclusion that evidence doesn't matter. One of the most compelling features of science is that it is self-corrective. As further inquiry generates new evidence, it enables researchers to recognize shortcomings of previous hypotheses and develop new ones that are better supported by that evidence. In this section, we focus on the strategies by which neuroscientists have gathered evidence about the nervous system. We will keep an eye both on how they enable researchers to learn and how they can sometimes lead researchers astray.

A major challenge in most sciences, including neuroscience, is that the phenomena about which we seek knowledge are not directly observable. Instead, researchers must rely on indirect evidence. When one opens up the skull to observe the brain, what one sees seems to be an inert object. In fact, there is a tremendous amount of physical movement occurring within the brain. Within individual neurons, what are called *molecular motors* (kinesins, dyneins, and myosins) are ferrying protein complexes and whole organelles to locations where they are needed. But these can only be observed with high-powered microscopes together with dyes that tag cargo being transported. These tools mediate our knowledge, and their reliability must in turn be established.

One form of knowledge about the brain that neuroscientists seek concerns the structural components of the brain – neurons, ganglia and nuclei, and laminar sheets. In the previous section, we saw how microscopes and techniques such as staining contributed to this knowledge. The knowledge sought, though, involved more than structure. Researchers elicited evidence that neurons transmit electricity along their membranes and communicate across synapses using neurotransmitters (Section 2.1). Establishing this information required techniques to measure activity and often to manipulate it. In this section, we introduce and examine some of the most prominent techniques that neuroscientists use to determine what brain components do.

A challenge in establishing what brain parts do is well illustrated by an approach that is now universally regarded as generating false claims. Franz Josef Gall (1812) hypothesized that the size of a brain region would correspond to how developed a trait was in a person. He further proposed that one could ascertain the size of regions in the neocortex from the contours of the scalp –

a bump on the scalp would correspond to the underlying region being abnormally expanded and an indentation to it being undersized. He then proposed to correlate these differences detected on the scalp with cognitive and personality traits, creating phrenological charts for cognitive and personality characteristics (Figure 8). Gall's assessment of correlation was selective and impressionist. This is not surprising, since the modern science of statistics would not be developed until the end of the nineteenth century. Some of his assumptions are demonstrably false: the scalp does not reveal the size of underlying brain regions, as there is space between them. Given his anatomical skill, Gall should have recognized this. Of far more interest is his central claim – that the size of brain regions corresponds to the development of a cognitive or personality trait. This is false, but not obviously so. Indeed, we often think that more of

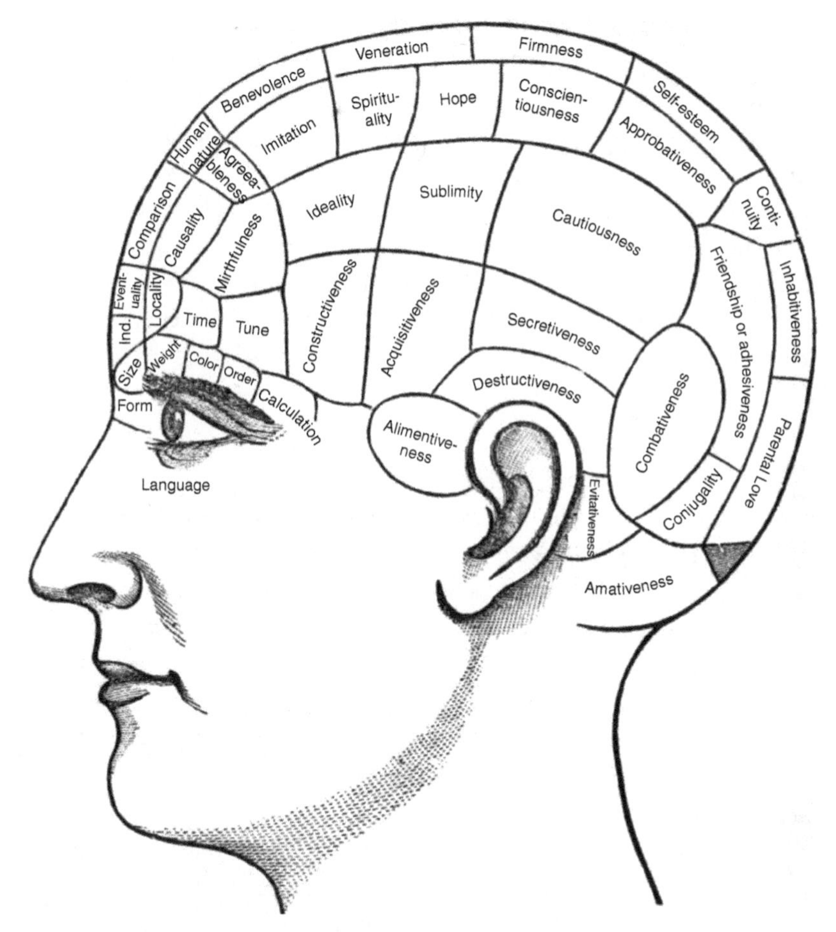

Figure 8 A phrenological map of cognitive and emotional capacities. Taken from Fowler (1890).

something should lead to more output. Neuroimaging techniques discussed later make a similar assumption in treating the amount of blood flow as a correlate of the amount of activity in a brain region.

Gall's approach, known as phrenology, attracted a great deal of popular interest throughout the nineteenth century. Scientists, however, were highly skeptical. Pierre Flourens undertook experiments that he believed refuted Gall. When he cut out (lesioned) neocortical regions in various animals, he did not find deficits in specific traits but only a general diminishment of mental capacities, with the degree of diminishment corresponding to the amount of cortex removed. He argued that this showed that different mental capacities were not localized in different parts of the brain.

As flawed as Gall's approach was, it reveals the basic strategy for developing and evaluating hypotheses about the functions of brain areas: find some means of relating values on a variable characterizing the brain area and values on a variable describing its hypothesized effect (or, as we will see in Section 3.3, its hypothesized cause). In the decades after Gall's endeavors, two such approaches gained traction among researchers: relating naturally occurring or experimentally induced damage (lesions) to brain regions with behavioral deficits and relating the electrical stimulation of a brain region with a measure of a behavior. We describe these in Sections 3.1 and 3.2, before turning in Section 3.3 to another approach that became the workhorse in the twentieth century: recording from brain areas either as sensory stimuli were presented or the person performed an activity.

3.1 Lesion Studies

A couple of decades after Gall, Paul Broca was brought in to oversee the treatment of a patient who had lost the ability to produce articulate speech (the patient is often referred to as Tan, after the one speech sound he could make). Broca made a bold prediction: damage in Tan's brain would be centered on the third frontal convolution. This prediction was vindicated on autopsy (Broca, 1861), and the region is now commonly referred to as *Broca's area*. Broca's research reveals an important challenge in lesion research – specifying what activity the damaged area performs in the undamaged brain. Broca characterized the area as responsible for articulate speech, but subsequent researchers viewed articulate speech as requiring the activity of many brain areas. The lesioned area may be needed for an activity but not capable of performing the activity on its own. Moreover, more capacities may be lost than initially suspected. Broca assumed that the area damaged in Tan had no relevance for comprehension, as Tan was able to comprehend what was said to him. More recently, however, researchers found that patients with damage to Broca's area have deficits in comprehending particular types of words, such as *on* or *under*, that signal grammatical relations. Over 150 years later, there is still considerable controversy about how to characterize the processing in Broca's area.

Another challenge with lesions in the human brain is that they often result from injuries or strokes and are not limited to the specific regions in which researchers are interested. In animals, researchers can try to target specific brain regions for destruction; however, they still face a major challenge – the brain is dynamic and often undergoes large-scale change after lesions in a specific area. The deficit manifest after the brain is lesioned may therefore not provide a very reliable indication of how the brain would have functioned with just the lesioned area removed. Researchers have developed a relatively new approach, known as transcranial magnetic stimulation, that addresses this shortcoming. Positioning a powerful magnetic coil next to the skull alters the electrical current in the underlying neocortex and can be used to temporarily impede processing in a region without allowing time for the brain to adapt (and without causing permanent damage to the experimental subject). The challenge still remains to determine what is the normal activity of the impacted area.

3.2 Stimulation Studies

Applying a stimulus to a brain region, typically through electrodes inserted into the brain, represents a second strategy. If such stimulation yields a detectable increase in a behavior, researchers infer that the brain area stimulated is responsible for the behavior. Cushing (1909) showed that a similar approach could reveal areas involved in sensory processing: after applying very weak electrical stimulation to primary sensory areas, people reported tingling sensations in different parts of their body. Further deploying this approach, Penfield and Rasmussen (1950) produced their famous homunculi images of the primary sensory and motor cortices (Figure 9). One notable feature of these images is that the areas for the mouth and hand are much larger than those for other body parts, which is interpreted as reflecting responsiveness to stimulation of these parts of the body and greater motor control over them. As with lesion studies, there are challenges in interpreting stimulation studies. Electrical currents can disseminate beyond the stimulated area and affect other processing. The behavior measured may reflect activity not only in the area stimulated but other areas to which the current is transmitted.

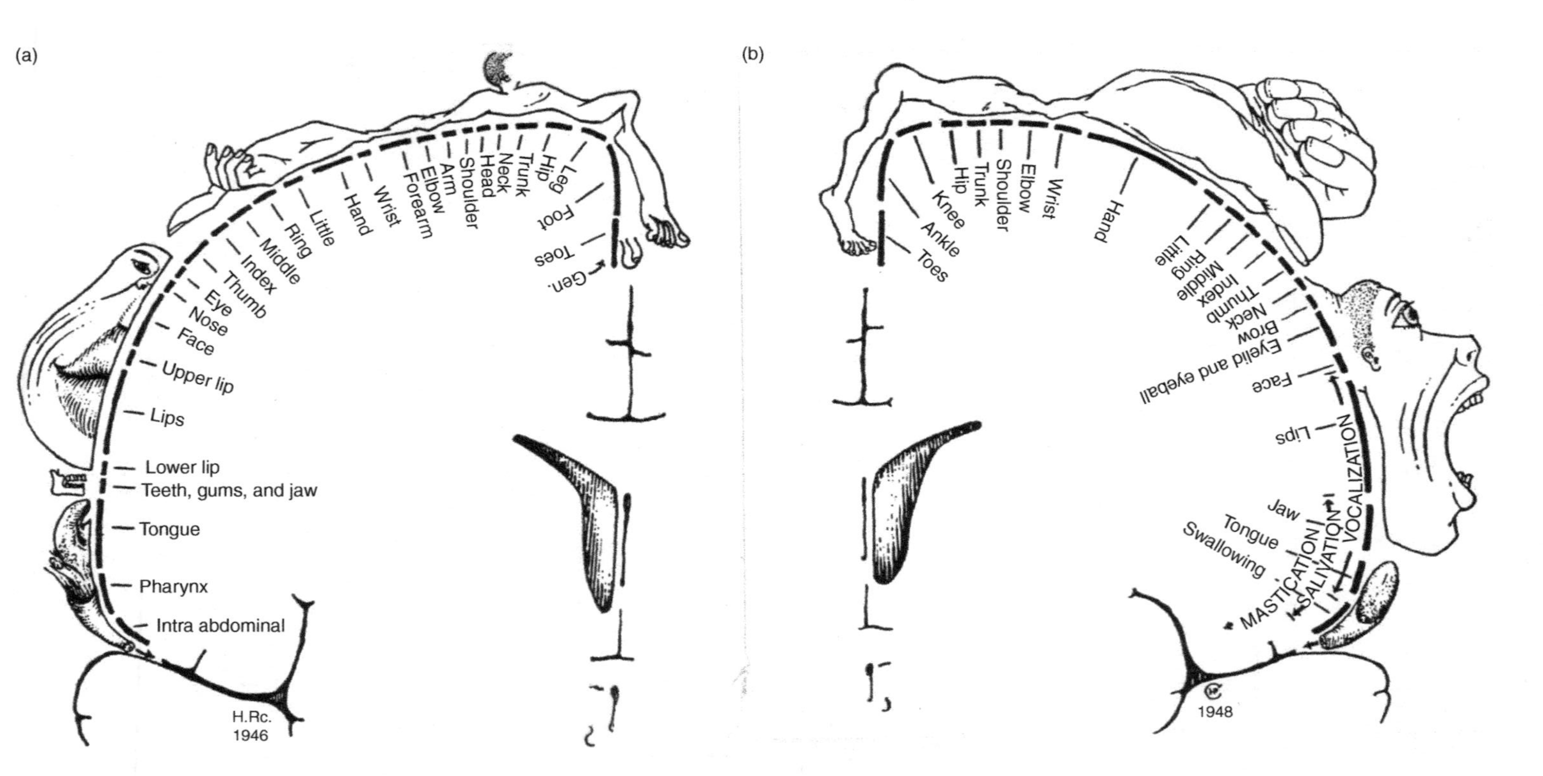

Figure 9 (a) Homunculi used to indicate areas in the somatosensory cortex (BA 1, 2, and 3) responsive to stimulation body regions and (b) to indicate areas in the primary motor cortex (BA 4) that activate body regions. Taken from Penfield and Rasmussen (1950).

3.3 Recording Studies

The approaches described so far involve altering brain activity and measuring effects on behavior. A different strategy is to record from brain regions as the organism is engaged in different tasks and infer the hypothesized causes of the brain activity. The initial application of this strategy involved inserting electrodes into the brain and recording the electrical activity generated by the neuron or neurons closest to the electrode. This approach, known as *single-cell recording*, became the workhorse technique of mid–twentieth century neuroscience. A powerful illustration of the approach was Hubel and Wiesel's (1959) exploration of neurons in BA 17, otherwise known as primary visual cortex or V1. By varying the stimulus presented to a neuron in a cat or a monkey while recording from it, they determined that edges elicited the largest response, with different neurons responding to edges at different orientations. Some cells responded when the edges were stationary, others to edges moving in specific directions. Inspired by these findings, researchers undertook numerous studies seeking to identify the features of stimuli that would elicit activity in different visual processing areas, as we will describe further in Section 5.3.

Inserting electrodes into the brain is highly invasive and regarded as morally unacceptable in humans except for patients being evaluated for neurosurgery to remove tumors. In that context, recording from neurons while patients perform tasks allows surgeons to avoid cutting areas regarded as of critical cognitive importance, such as those involved in producing or understanding language. Researchers have often been able to take advantage of these situations and, with the patient's permission, record from neurons as different stimuli are presented. In a widely cited study, Quiroga et al. (2005) identified neurons that responded selectively to different images of well-known people, such as Jennifer Aniston or Bill Clinton, or even to their spoken name (suggesting that the neuron was responding to the person, not their visual appearance).

Given the moral issues in inserting electrodes in order to record brain activity, researchers working on humans often turn to noninvasive techniques for recording neural activity. One of the first to be developed was the electroencephalogram (EEG), which records electrical signals from electrodes placed on the scalp. In the pioneering research with this approach, Berger (1930) detected oscillations that varied in frequency depending on the activity that the subject was performing. When participants were quiet and kept their eyes closed, he detected oscillations of approximately 10 Hz (he referred to these as *alpha waves*). When participants opened their eyes or were asked to perform a cognitive task, the rhythms would increase to between 20 and 30 Hz and the amplitude would decrease (he called these

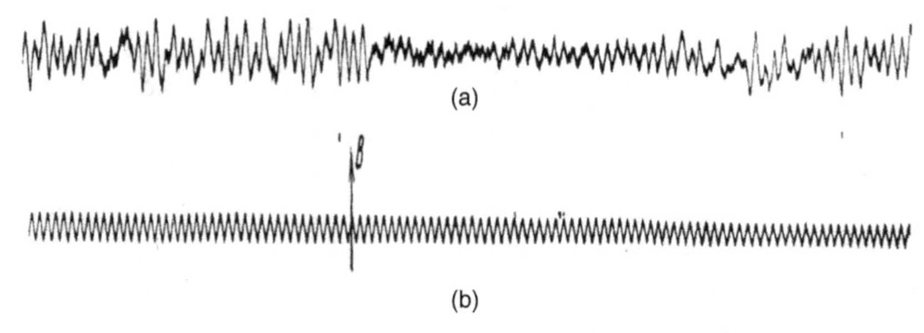

Figure 10 (a) An eight-second recording of a subcutaneous EEG by Berger (1930). (b) A 10 Hz timing signal. Initially, the EEG consists of alpha waves, but shortly after Berger strokes the subject's hand with a glass rod (indicated by the B on the timing bar), a period of low-amplitude, shorter-period beta waves ensues, which then transitions back to alpha waves.

beta waves) (Figure 10). In ensuing years, researchers differentiated yet faster and smaller amplitude gamma rhythms (above 30 Hz), which appear during cognitively demanding tasks, and slower delta (less than 4 Hz) and theta (4–7 Hz) rhythms, associated with the transition to sleep. The current measured and recorded in these oscillations is not due to action potentials but to synchronized fluctuations in the resting potential of a population of neurons (see Section 2.1). EEG has attracted renewed interest as researchers have recognized the importance of oscillations in cortical processing (see Section 9.3).

While EEG can measure brain activity with high temporal resolution, its spatial resolution is very poor, in part because electrical activity disperses widely. Noninvasive techniques that measure blood flow as a proxy for neural activity, such as functional magnetic resonance imaging (fMRI), offer much higher spatial resolution (albeit with a loss of temporal resolution since the rate of blood flow changes slowly). To interpret the neural activity in terms of contributions to behavior, researchers ask subjects to perform different cognitive tasks while lying in a scanner. When there is a large change in blood flow, researchers infer that this particular brain region is selectively contributing to the activity. The challenge is to infer just what an area has contributed to a task. Often this is addressed by exploring different tasks and asking what feature the tasks that elicit greater activity have in common. The challenge of identifying the contribution of a given area is increased when not just one but multiple brain regions exhibit increased activity in a given task. One response has been to shift the focus from individual brain regions to identifying networks of brain areas whose activity changes in a coordinated fashion. The goal is then to correlate these networks with the type of processing required in the tasks that activate these areas (see Section 6.4 for analysis of networks).

Marcus Raichle, one of the pioneers in fMRI studies, drew attention to the fact that while some brain regions increase their activity when a subject is performing a task, others exhibit reduced activity. Raichle et al. (2001) scanned subjects when they were not assigned a task (a condition referred to as the *resting state*) and identified a network of brain regions whose activity was heightened in this condition and reduced in task conditions. Raichle termed this the *default mode network*. This opened up a new line of inquiry into what brain regions are doing when a person is not specifically challenged to perform a task.

3.4 Altogether Now

Since each of the methods we have reviewed – lesion, stimulation, and recording – present different epistemic challenges, one strategy for reducing these challenges is to employ all three together. We illustrate this approach with research on an area in BA 19 variously known as V5 or MT. Human patients with lesions in this area are unable to detect motion (rather, they see the world as a sequence of still images). To explore how V5 contributed to detecting motion, Britten et al. (1992) presented monkeys with displays in which objects were either moving in a common direction or in random directions and trained them with food rewards to press a different lever for each condition. They then introduced an ambiguous display in which about half of the objects moved in a common direction, while the rest moved randomly. The monkeys still responded, sometimes selecting the direction in which half the dots were moving and sometimes selecting the lever that indicated that they were moving randomly. By recording from neurons in MT, the researchers identified neurons that corresponded to the response that the monkeys made to the ambiguous display and inferred that activity of these neurons constituted the monkeys' perception of motion. Finally, they microstimulated neurons that corresponded to a specific response, and showed that they could bias the monkeys to make that response. They took these findings as strong confirmation that these neurons are responsible for monkeys' perception of motion. In this research, results from lesion, stimulation, and recording studies all converged, making the case especially compelling. Although one might challenge each of the techniques alone, using them together provides strong evidence that area MT is involved in visual perception of motion.

Underlying these efforts to determine the activity performed by a brain area is the idea that brain areas are modules responsible for specific tasks – for example, MT is an area for detecting motion. As compelling as the case that MT processes motion information is, subsequent research has shown that this is not all MT does. It is also active in binocular vision, for example. Drawing on such evidence, Burnston (2016) and Anderson (2014) argue against fixed assignments of specialized functions to brain areas, arguing instead for more contextualized accounts in which brain regions perform multiple types of processing. What they do in a given situation depends on features of that situation: in different situations, they form coalitions with different brain areas and perform different functions. If such claims are correct, they can explain the versatility and adaptability of neural processing, but they make the challenge of figuring out what brain areas do much more difficult.

3.5 Computational Modeling

We finish by briefly noting another method utilized in neuroscience – computational modeling. This involves identifying variables thought to describe the changing states of a neural system and developing equations (typically differential equations) that characterize how values of these variables change as the values of other variables change. By starting with values for variables that are thought to describe the nervous system at one point in time, and having a computer iteratively apply these equations to determine subsequent values, researchers seek to simulate the brain system. A successful stimulation is one that generates a succession of values of variables that correspond to those measured in the brain. When the equations formalize what is already hypothesized about how the brain functions, a successful simulation provides evidence that the hypothesized account is correct (also see Section 6.3 on the contribution of computational modeling to neuroscientific explanation).

Computational models can also be developed as part of a discovery process. If one succeeds in developing a simulation that matches the behavior of the brain, one can interpret the equations as hypotheses about the processes actually operative in the brain. In that case, though, one generally seeks independent evidence that there are processes in the brain that correspond to the equations.

3.6 Summary

To investigate brains, researchers use a variety of techniques including lesioning, stimulating, and recording. As we have noted, the inference from these studies to what is happening in the brain is often indirect. Accordingly, researchers must often combine multiple methods. Brain researchers also often invoke computational models, which allow them to determine how the brain would work if a particular hypothesis was correct or, in some cases, to advance new hypotheses.

4 From Whom Do Neuroscientists Learn about the Nervous System?

To learn about nervous systems, researchers must actually study nervous systems, using methods such as those introduced in the previous section. But whose nervous systems should they study? If the researcher is interested in a specific individual, then they would reasonably choose to study that individual. But science is generally focused on types, not tokens, where types are classes of entities taken to be the same in relevant respects. The goal is to generalize across the members of the type. This is relatively straightforward in the physical sciences. Chemists are not interested in a given specimen of, for example, gold, but in all instances of gold. What they discover in studying one specimen is assumed to apply to all instances. Neural scientists seek similar generalizability, although the scope of generalization is less clear cut.

A variety of characteristics can be used to identify types of organisms. For example, one might be interested in left-handed human beings. One might focus on species, for example, humans. Species membership is not determined in terms of necessary and sufficient conditions, as it is with elements like gold. Instead, what is relevant is the organism's history: Who were its parents? As species themselves originate from other species (as members of a person's family arise from other members of the person's family), these relations are often represented in branching trees. These descent relations correspond to inheritance – genetically based traits that emerge at one node in the tree are generally inherited by the branches. In this respect, evolution is a conservative process: as observed by Ernst von Baer in the decades before Darwin published his account of evolution through natural selection, new traits develop as variations and modifications of existing traits. Accordingly, generalization in biology, including neuroscience, involves applying what is learned about some species to those appearing in a particular clade (descendants of a common ancestor) in the evolutionary tree. One common way in which variation arises in descendants is with a mutation in which part of a chromosome is duplicated, generating multiple copies of some genes. Through further mutations coupled with natural selection, these duplicated genes differentiate and code for proteins that perform specialized tasks. As a result, descendant species retain the same basic traits but give rise to specialized versions. For this reason, biologists often find it useful to look back in the evolutionary tree to where a trait first emerged.

They can then study the traits in the simpler organism that gave rise to those in later organisms.

In this section, we start with the challenges of doing research on people and then turn to research on model organisms that are assumed to reveal many of the same traits as those in the organism of primary interest (typically humans) due to sharing a common ancestor.

4.1 People

Since we are generally most interested in the human brain, it makes sense to study people. The problem is that it is morally objectionable to use many techniques discussed in the previous section with humans. Although medical research historically was done on individuals without their consent, we now require any participant in research (or, in some cases, their proxy) to give informed consent (there is considerable debate about what actually constitutes being informed or genuinely consenting). However, even if someone was willing to allow invasive research techniques to be applied to them, society has judged this to be unacceptable. We do not allow people to consent to having parts of their brains removed or to have electrodes inserted into their brains except when it is judged to be therapeutic (in some cases, as in the example discussed in Section 3.3, an individual can consent to participating in additional research when the invasion is required for therapeutic ends).

Far more common is research on individuals with brain damage. As with the example of Broca's patient Tan, researchers hope to gain insight into what the damaged area contributes in individuals without damage (see Section 3.1). Sometimes damage to the brain is produced by accidents. This was the case with Phineas Gage – in an accident during railroad construction, a large iron rod was driven through much of his left frontal lobe. The accident resulted in major personality changes that were reported by his friends and caregivers, and drawing on these reports and studies of patients with similar injuries, Damasio (1995) has argued that the areas damaged in Gage are involved in employing emotions in making decisions. In some cases, when the individual whose brain is damaged gives consent, researchers can deploy a host of tests to determine the effects of the brain damage. For example, after he underwent surgery to remove his hippocampus in the attempt to treat epilepsy, Henry Molaison (often referred to as HM) lost the capacity to develop new memories of events in his life. He became the focus of numerous studies directed at determining both which capacities he had lost and which he had retained (Corkin, 2013).

Until recently, the only way to study brains of healthy individuals was to examine their behavior. By contrasting the behaviors a person could perform and those they could not, for example, researchers could draw inferences about how their brains must be organized. Increasingly, noninvasive techniques such as EEG and fMRI (Section 3.3) enable researchers to record activity in brains. Although there techniques have revealed much about human brain activity, they have serious limitations. Consider trying to figure out how a car engine works from using listening devices to record the activities occurring in it as it functioned normally. Both with cars and brains, a great deal of reasoning is required to infer from these external measurements what is happening inside.

4.2 Model Organisms

Since our society places fewer constraints on what can be done to members of other species, neuroscientists perform much more research on nonhuman animals.[3] Some species have been selected for research and are considered *model organisms* (Ankeny & Leonelli, 2020). In some cases, researchers choose to work on a given species because of the relative ease of obtaining results that are clear and easy to interpret (we saw examples in Section 2.3 in which leeches were selected, and in Section 3.3 in which cats and macaque monkeys were selected). In other cases, researchers choose to work on an organism because techniques have already been developed to study it, and data have already been amassed against which to evaluate results. This accounts for extensive use of fruit flies and mice (in both cases, researchers can procure animals from companies that breed pure strains, reducing variability that interferes with interpreting results).

As we noted earlier, since evolution is conservative, researchers often prefer to investigate organisms that are thought to resemble ancestral organisms in which the trait of interest first appeared. Increasingly, researchers are looking to bacteria and plants as model organisms for understanding behavior and cognitive activities, but here we limit ourselves to animals.

The value of investigating distantly related organisms is illustrated in research on sleep, which remains one of the most puzzling features of animal behavior (sleep renders an organism vulnerable to predators for prolonged periods). Until 2000, most sleep research was performed on humans or other mammals (e.g., rodents). In that year, two research groups showed that fruit flies

[3] There have long been individuals who oppose research on some or even all animals, but there has not been a consensus to stop all animal research. Today, there are strong prohibitions against doing invasive research on higher primates such as chimpanzees, and rules have been developed for the care of other species when they are used in invasive research. Debates continue about whether to permit invasive research and, if so, what sort.

exhibit the behavioral traits of sleep – quiescence in a stereotypic posture, requiring a stronger than normal stimulation for arousal, and a need to make up for lost sleep (Hendricks et al., 2000; Shaw et al., 2000). This opened up a new opportunity to understand sleep. Fruit flies manifest fewer gene duplications (discussed earlier) and so provide a vista into the basic mechanism. Moreover, a rich set of experimental procedures have been developed to investigate fruit flies. Drawing on these, researchers have made progress in understanding the molecular processes involved in sleep (Joiner, 2016).

To further illustrate model organism research in neuroscience, we turn to *Caenorhabditis elegans*, a small (about 1 mm in length) free-living, transparent round worm with a life cycle of less than three days. One reason it is often selected for research is that it is the only organism for which researchers have an almost complete map of its nervous system. This was generated through laborious research that required slicing a worm thinly, making electron micrographs of all the slices, identifying the neurons and their projections in each slice, tracing them to adjacent slices, and then piecing the results back together (White et al., 1986). The resulting map (referred to as a *connectome*) was consistent across hermaphrodite worms and included 302 neurons (each named with three letters) and approximately 1,000 projections between them.

This relatively simple nervous system raised the prospects of determining how neurons controlled the worm's behavior by identifying circuits within the network responsible for specific behaviors.[4] Chalfie et al. (1985) identified a relatively simple neural circuit that controls a withdrawal response in response to touch – if the worm is touched in the head region, it reverses its movement, whereas if touched in the rear, it accelerates (Figure 11). The circuit is relatively easy to understand. For example, the sensory neuron activated by touch to the tail, PLM, is connected by gap junctions to PVC, which sends excitatory connections to motor neurons that generate forward motion. PLM also inhibits AVD (via a chemical synapse), thereby inhibiting backward motion. In a similar way, anterior touch results in backward movement.

The simplicity of the withdrawal response circuit affords understanding, but that does not mean that there are not complications, some of which have been revealed by further research. First, the circuit is modified by learning (Ardiel & Rankin, 2010). The worm exhibits what is referred to as *short-term habituation* – if it is tapped repeatedly but experiences no adverse effects, the worm responds less frequently and withdraws less distance. This is not just due to the worm becoming tired: the degree of reduction increases when the interval between touches is longer,

[4] Among the behaviors it exhibits is sleep-like behavior during stages of its development, rendering it a model organism for sleep studies as well (Keene & Duboue, 2018).

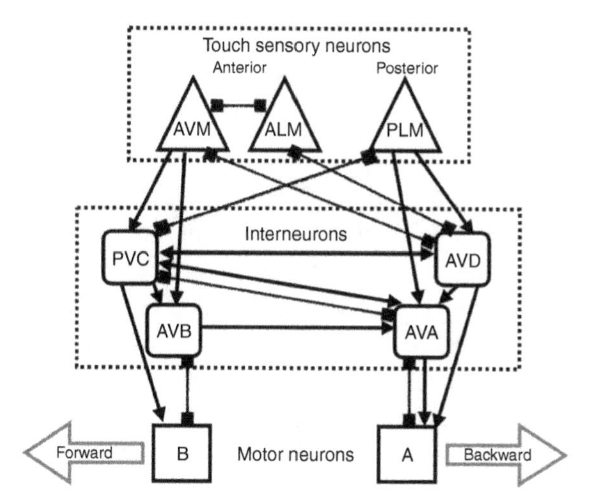

Figure 11 Touch withdrawal response network in *C. elegans*. Arrows between neurons represent chemical synapses, while box-ended lines represent gap junctions.

the opposite of what we would expect if the effect were due to fatigue. It also exhibits long-term habituation in that its behavioral change can last more than a day (more than a third of the *C. elegans'* life cycle). The effects of long-term habituation exhibit similarities to those of human learning: habituation is stronger if exposure to stimuli is spaced out[5] and requires active relearning after recall.[6] The basic processes involved in long-term habituation have been identified (increased receptors for the neural transmitter glutamate in the neurons intermediate between sensory and motor neurons) and present an attractive way to study memory mechanisms in humans.

The withdrawal response circuit is also subject to modulation by volume transmitters such as dopamine and serotonin (introduced in Section 2.3). Worms without food exhibit more habituation to taps, an effect that is mimicked by application of dopamine, suggesting that endogenously released dopamine modulates habituation when food is lacking. Ardiel and Rankin (2010) and Bargmann and Marder (2013) described numerous other control circuits for specific behaviors in *C. elegans* that are altered by neuromodulators (for

[5] You have no doubt been told that cramming for exams is not a good way to learn. Spacing out learning episodes leads to improved learning in us and better habituation in worms.

[6] Human memory researchers have determined that, after exposure to an event, a process of consolidation is required for the memory to endure. After recalling a memory, a similar process, referred to as reconsolidation, is required (administration of a memory blocker during consolidation or reconsolidation will eradicate the memory). The requirement that habituating worms also must relearn after recall suggests that they employ similar processes of consolidation and reconsolidation.

philosophical analysis of this research, see Anderson, 2014). The relatively simple circuits in *C. elegans* both render detailed study possible and also reveal how such circuits can be modulated to generate complex behavior.

4.3 Summary

Much of the interest in neuroscience stems from a desire to learn about the human nervous system. In some cases, such as when an accident produces damage to a person's brain or when the detection of brain activity does not require intrusion into the brain, researchers can study humans. But in many cases, they cannot. Moreover, there are often advantages to working with nonhuman species. In some cases, researchers elect to investigate simpler organisms as it is often easier to figure out the basic principles by which the nervous system works.

5 What Has Neuroscience Learned?

In previous sections we introduced some of the tools used in neuroscience and the organisms that it investigates. At this point it will be helpful to introduce some examples of what neuroscience has learned about vertebrate brains, including our own. We will make use of these examples in subsequent sections as we engage in philosophical discussions about neuroscience.

5.1 Keeping Track of Time of Day in the Suprachiasmatic Nucleus

We start with a nucleus within the hypothalamus (labeled in Figure 1), the suprachiasmatic nucleus (SCN). The hypothalamus is a collection of nuclei that play critical roles in regulating fundamental activities such as eating, maintaining wakefulness or going to sleep, and reproduction (Leng, 2018). Individual nuclei receive inputs and send outputs to various regions of the body but also to regions elsewhere in the brain. Most release neuropeptides and volume transmitters that diffuse broadly, modulating activity of other neurons as well as controlling physiological processes. For example, the arcuate nucleus contains neurons that respond to peptides released in the intestinal tract that signal fat concentration or whether food is being digested. The outputs of these neurons in turn regulate eating behavior (Sohn, 2015). A nucleus in the lateral zone of the hypothalamus contains hypocretin-producing neurons that play critical roles in maintaining wakefulness or transitioning to sleep – silencing these neurons induces slow-wave sleep (Burk & Fadel, 2019). The role of the hypothalamus is sometimes minimized as merely engaged in bodily maintenance. A useful corrective is to reflect on how much of our behavior is focused on activities such as eating, sleeping, and reproducing.

Coordinating our activities with the light-dark cycle on our planet is of fundamental importance. Although artificial lighting allows us to carry on our activities around the clock, our physiological and cognitive activities are affected by endogenously produced rhythms of approximately 24 hours (named, *circadian* from *circa*, approximately, + *dies*, day). These become apparent to us as jetlag when we travel across multiple time zones, but they also manifest in the increased rates of obesity, cancer, and other conditions in shift workers. Enzymes responsible for the activities of nearly every organ in our bodies exhibit oscillating expression over the course of a day, thereby resulting in varying performance. This includes regions of the brain that are involved in higher cognitive activities such as reasoning and decision-making. Our capacity to perform these activities varies over the course of the day.

Research on fruit flies provided the first clues to how circadian rhythms are generated.[7] A search of genetic mutations revealed one gene, named *period* (or *per*), in which mutations altered the period of rhythms or eliminated them altogether (Konopka & Benzer, 1971). Genes are transcribed into messenger RNAs (mRNAs) and translated into proteins. Still working in fruit flies, Hardin, Hall, and Rosbash (1990) established that concentrations of both the *per* mRNA and the protein Per oscillated over a 24-hour period, with the protein lagging a few hours behind the mRNA. Since a negative feedback loop is a common mechanism for generating oscillations (see Section 6.3 for further discussion), they proposed that rhythms resulted from a feedback process: as the protein Per accumulated, it inhibited the expression of the *per* gene, resulting in the concentration of the protein subsequently diminishing, only to increase again when Per itself degraded.

Studies lesioning the SCN or recording from SCN neurons revealed that circadian rhythms are generated in much the same manner in vertebrates (Takahashi, 2017). While the oscillations occur within individual SCN neurons, the connections between them turn out to be important. Individual neurons generate rhythms with different periods; only as a result of each neuron modulating the activity of others does a regular oscillation of approximately 24 hours arise (Welsh, Takahashi, & Kay, 2010). As that oscillation is still only approximately 24 hours, the SCN, like an old-fashioned watch, will drift gradually from the correct time and frequently has to be reset by external cues. The effects of not doing so were shown in classic experiments in which humans lived in enclosures without external time cues. Their circadian rhythms were somewhat longer than 24 hours, leading them to rise several minutes later each day.

[7] Fruit flies do not have an SCN; rhythms are instead maintained by a small collection of neurons in their brain.

Keeping the oscillations in the SCN synchronized with the day–night cycle on our planet is achieved through daylight, which in humans is only processed by the eyes.[8] Daylight also plays a central role in the human ability to overcome the effects of jetlag. One of the best ways to adjust to a new time zone is to time exposure to daylight appropriately.[9] Conversely, avoiding daylight after night-shift work can minimize the ill effects of rotating shiftwork, which stem from the constant resetting of the clock as a result of continuing changes in timing of light exposure.

5.2 Mapping Location in Space in the Hippocampus

Knowing one's location in space and how to navigate to other locations is an extremely important behavioral capacity. Animals often depend on being able to figure out routes to food or their nests. Tolman (1948) demonstrated this ability in rats running mazes. When a new, more efficient route became available or when the route previously used was blocked, rats flexibly adjusted the routes that they took to a food location. He inferred that the rats were using *cognitive maps* to determine their routes much in the manner humans use physical maps. Tolman, however, did not have the research tools to identify where these maps are in the brain and how the rats could use them to guide behavior.

Clues to the location of cognitive maps in the brain came from research on deficits in rats with damage to the hippocampus, a cortical structure adjacent to the temporal lobe of the neocortex. In experiments using the Morris Water Maze – an apparatus in which rats are forced to swim until they find a hidden platform on which they can stand – normal rats learn the location quickly and swim directly to the platform from wherever they are in the maze. Rats with damage to the hippocampus fail to learn, continuing to swim erratically on subsequent trials (O'Keefe & Nadel, 1978).[10] Little was known at the time

[8] In other animals such as birds, the pineal gland, which Descartes thought was the locus of interactions between the mind and body, is sensitive to light that passes through the skull. Although it is not the locus of mind–brain interactions, the pineal gland is important in mammals, including us: it releases melatonin, which serves to reset the SCN neurons (hence, the popular use of melatonin to counter jetlag).

[9] The key is to target exposure to the period just prior to normal first light exposure in the location from which one started when traveling eastbound. Since the clock can only adjust about one hour per day, it is important to advance this exposure about an hour each day.

[10] During the same period, a surgery to remove the hippocampus in order to reduce epileptic seizures in Henry Molaison (HM; see Section 4.1) resulted in his total inability to develop new explicit memories, leading to the hypothesis that the hippocampus is the locus in which new explicit memories are initially encoded. Since HM could remember events from years before the surgery, researchers have proposed that over time memories are transferred to areas in the neocortex. Initially the rodent researchers and the human researchers viewed themselves as studying different phenomena. More recently, though, there have been attempts to connect the two abilities that are lost when the hippocampus is damaged (Eichenbaum, 2002).

about the function of hippocampus. The hippocampus had primarily been a source of neurons to study individual neuron behavior. Typically, neurons cease to respond to stimuli over time, but researchers discovered that if they applied a brief but intense sequence of electrical stimulations to hippocampal neurons, they would continue to respond. This phenomenon, now known as *long-term potentiation*, indicated that response properties of neurons could be altered and offered a model of how neurons in the hippocampus (and elsewhere) can quickly alter their response properties (a feature important for learning). (See Craver, 2003, for a philosophical examination of this history.)

To figure out how the hippocampus could constitute a cognitive map, John O'Keefe recorded from neurons in the hippocampus while a rodent occupied different locations in an enclosure. He found that different neurons would increase their firing rate when the rodent was in different specific regions of its enclosure. O'Keefe and Conway (1978) named these neurons *place cells*. Since different cells functioned as place cells in different environments, numerous researchers began to examine how place cells would respond as they morphed one environment into another. Some changes, such as rotating the enclosure or shrouding it with a curtain, did not affect the activity of place cells, but others, such as altering cue cards placed on the walls of the enclosure or significantly increasing the size of the enclosure, did. Studying how neurons altered their response patterns with environmental changes, such as gradual transformation of a square enclosure into a circular one, provided an avenue for studying how these maps are established (Colgin, Moser, & Moser, 2008).

In Section 3.3, we described how EEG identified ongoing oscillations of subthreshold electrical activity in many brain areas. These oscillations can also be detected intracranially. In the hippocampus there is an ongoing theta oscillation (6–12 Hz). Comparing the timing of action potentials in place cells with this oscillation, O'Keefe and Recce (1993) revealed that when the rat first entered the area to which a place cell would respond (its place field), the place cell would fire at the trough of the theta cycle. On each subsequent oscillation, it would fire at a slightly earlier phase of the theta cycle (Figure 12). Comparing the timing of action potentials in neurons for nearby regions provided the rodent a means to track its location along a path (place cells that fired earlier in the theta cycle represented locations arrived at earlier). An even more striking finding was that when rodents were allowed to run down a runway, the activity of place cells after they finished the run would reflect the sequence of place cell activity during the run, but in reverse. If the animals were delayed in starting the run, place cells would fire in the same sequence as when they subsequently ran (Diba & Buzsáki, 2007). This indicated that place cell activity serves to replay or anticipate future routes.

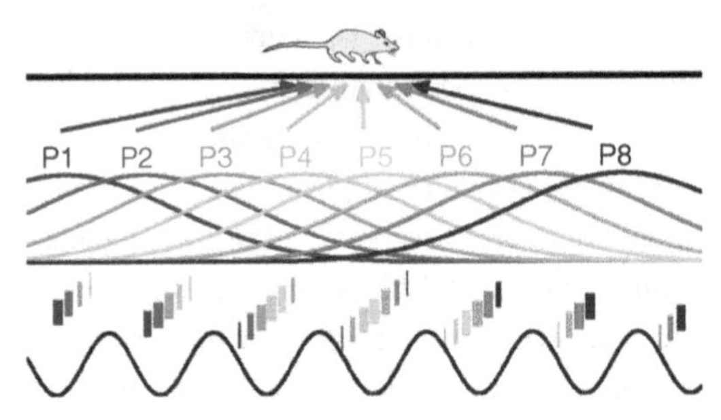

Figure 12 Neurons responding to place fields P1–P8 fire in relation to the ongoing theta cycle. Those representing the current location produce more spikes (indicated by the width of the bar) at the trough of the theta cycle. Those representing earlier locations issue fewer spikes and are earlier in the theta cycle. Those representing future locations also issue fewer spikes, but later in the theta cycle. Reprinted from Buzsáki (2010) with permission of Elsevier.

5.3 Seeing the World with the Visual Neocortex

In Section 3.3, we described how Hubel and Wiesel (1959) presented visual stimuli while recording from neurons in BA 17/V1 and determined that these neurons responded to edges in the visual scene. These researchers noted that detecting edges was only an early step in seeing the world and initiated a project of recording from areas in front of V1. They determined that neurons in a region of BA 18 that came to be known as V2 are able to respond to illusory contours (Figure 13). Moving progressively forward in the brain, researchers such as Semir Zeki (1971) showed that V4 neurons responded to shapes and that V5/ MT neurons, as we discussed in Section 3.4, respond to motion. (The adjacency relations between these areas are indicated in Figure 14A, a map developed to show, as well as possible, these relations in two dimensions.)

Based on the different effects of lesions to regions in the temporal and parietal lobes in monkeys, Mishkin, Ungerleider, and Macko (1983) advanced the hypothesis that visual processing proceeds along two separate pathways from the occipital cortex: a dorsal pathway to the parietal lobe in which neurons respond to information about where the stimulus is in the visual field (the *where* pathway), and a ventral pathway to the temporal lobe in which neurons respond to the identity of the object serving as the stimulus (the *what* pathway). Recording from neurons at the end of the what pathway in the inferotemporal cortex revealed that they respond to the identity of objects wherever they appear in the visual field. For example, Gross, Rocha-Miranda, and Bender (1972) identified neurons that responded to the human hand however it was oriented. Neuroimaging studies on humans revealed that one widely discussed area in the

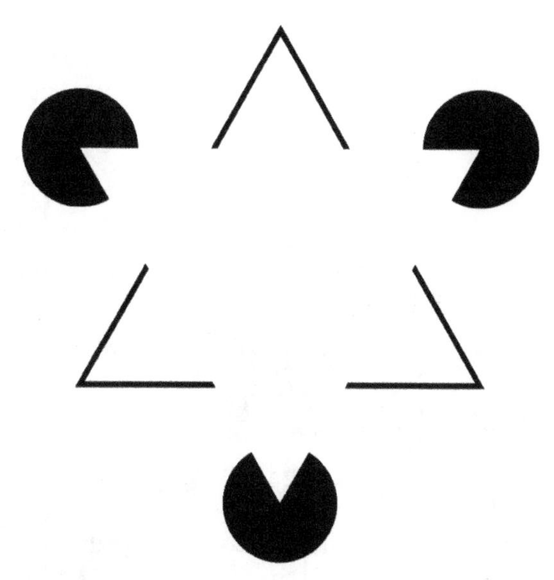

Figure 13 In this figure, originally developed by Kanizsa (1976), the white triangle appears to have boundaries, but these are illusory. Image by Fibonacci – own work, CC BY-SA 3.0, https://commons.wikimedia.org/w/index.php?curid=1788215.

pathway, the *fusiform face area*, responds to faces of specific people (Kanwisher, McDermott, & Chun, 1997). Individuals with lesions in this area experience prosopagnosia (the inability to recognize the faces of people they know). It is in this area that Quiroga et al. (2005) identified neurons that responded selectively to pictures of Jennifer Aniston or Bill Clinton (see Section 3.3). There is ongoing controversy as to whether these neurons respond only to faces or also to specific individuals in categories such as trees.

Neurons at the top of the where pathway in the posterior parietal cortex respond to location information but in terms of coordinates fixed by one's head (Andersen, Essick, & Siegel, 1985). Van Essen and Gallant (1994) synthesize much of the information about different visual processing regions into an account of two processing streams (Figure 14), referring to streams rather than pathways to recognize that there are various points of crossover from one stream to the other.

Accounts of just what is being processed in a brain area are often controversial. Milner and Goodale (1995) challenged the what/where distinction, arguing instead, on the basis of deficits in patients with brain damage, for a distinction between vision for perception and vision for action. One patient they studied, with temporal lobe damage, could see features of objects but was unable to

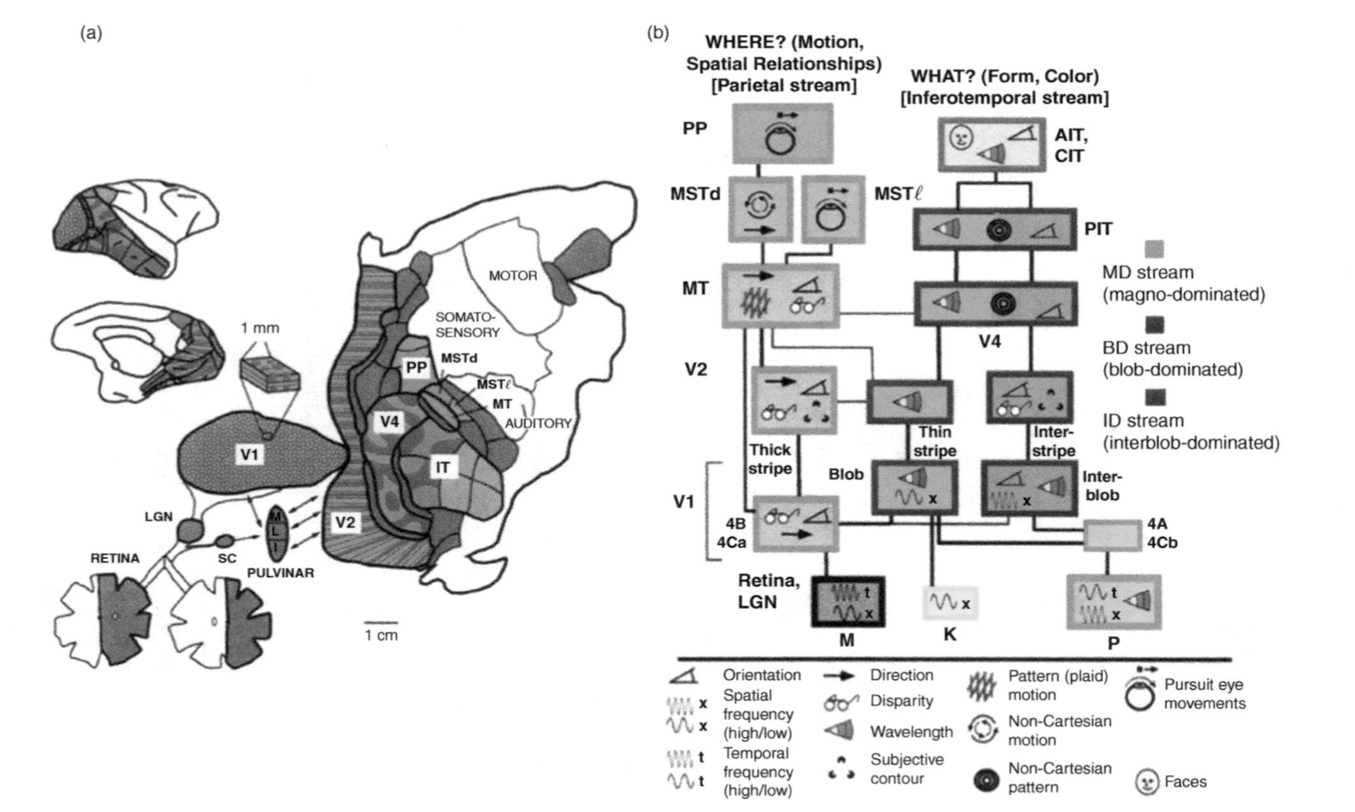

Figure 14 (a) A flat representation of the neocortex, with areas known to be involved in vision shown in color or dark gray. (b) A schematic diagram of the processing pathways involved in vision. Reprinted from van Essen and Gallant (1994) with permission from Elsevier.

recognize what type of object it was. She could only draw them laboriously. She also could not describe the orientation of a slot but nonetheless was able to correctly insert a letter into it. Goodale and Milner concluded that the areas spared in the parietal stream were not generically involved in processing location information but rather served to coordinate visual information with actions. The disagreement between Mishkin, Ungerleider, and Macko, and Milner and Goodale illustrate the challenge of settling what information a brain area is actually processing.

5.4 Making Decisions in the Basal Ganglia

Decision-making is a fundamental activity for all organisms as they can only perform some of the activities available to them at a given time. This is clear if one considers locomotion – if an organism moves, it cannot remain still. The network in *C. elegans* we described in Section 4.2 makes decisions between forward and backward movement. The basal ganglia, several interconnected nuclei found in all vertebrates, including those without a neocortex, play a particularly important role in decision.

The critical role the basal ganglia play in selecting behaviors is illustrated by mid–twentieth century research involving cats and other animals in which neural pathways were destroyed at various points between the CPGs (discussed in Section 2.2) and the neocortex (Figure 15). Stimulating the CPGs in organisms in which the CPGs were cut off from the rest of the brain resulted in specific movements that adjusted only to feedback from the muscles themselves. When lesions were between the mesencephalic locomotor region (MLR)/the diencephalon locomotor region (DLR) and the reticulospinal system, electrical stimuli to the reticulospinal system generated activity in multiple CPGs but not overall coherent activities such as walking or running. However, if the lesion is above the MLR/DLR, severing the connections with the thalamus, basal ganglia, and neocortex, stimulation of MRL neurons resulted in coherent movement patterns such as walking or running. The animals also avoided obstacles and elicited appropriate metabolic activity to meet the energy demands of muscle activity (Grillner & El Manira, 2019). What these results point to is a hierarchy of areas involved in ever larger-scale coordination of activity.

With lesions below the level of the basal ganglia and thalamus, animals only perform activities when stimulated; they do not initiate them. If the thalamus and basal ganglia are spared, however, animals are able to initiate and choose between activities. For example, cats whose neocortex was removed in infancy but retained an intact thalamus and basal ganglia walk, explore their

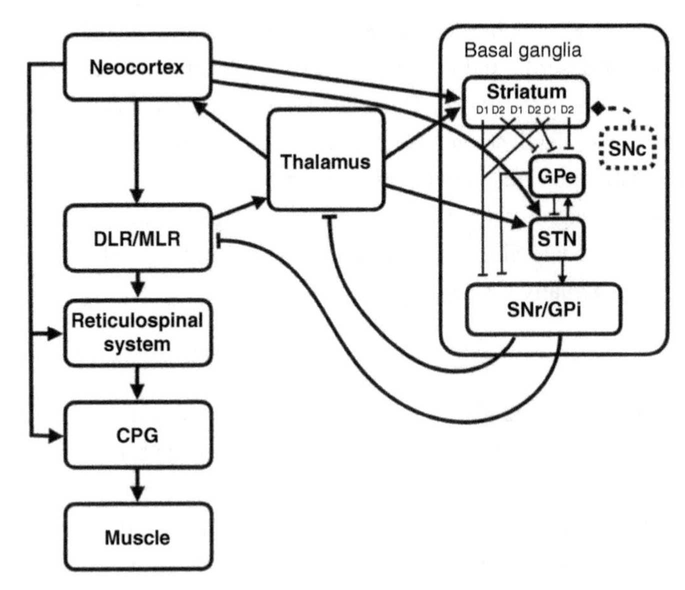

Figure 15 Neural centers operating on muscles. Arrows represent excitation, edge-ended lines inhibition. The dotted line with a diamond end indicates dopamine modulation for D1 and D2 neurons in the striatum.

environment, clean themselves, seek and eat food, and so on. They can live and fend for themselves for several years in the protected environment of the laboratory (Bjursten, Norrsell, & Norrsell, 1976). To do this, they must regularly make decisions about which actions to perform.

The structure of the basal ganglia suggests its role in decision-making. As shown in the upper right of Figure 15, the basal ganglia are an interconnected collection of subcortical nuclei. It is easiest to understand how they function by starting with the output nuclei: the substantia nigra pars reticulata (SNr) and the globus pallidus internal (GPi). Projections from neurons in these areas to other parts of the brain are all inhibitory. Moreover, they generate action potentials without external stimulation. Hence, by default, the basal ganglia inhibit or shut down activity elsewhere in the brain. What processing elsewhere in the basal ganglia does is selectively remove this inhibition, thereby allowing target areas to perform their activities.

Decisions as to whether to release other brain regions from inhibition are made through the collective operation of three pathways within the basal ganglia: the direct, indirect, and hyperdirect. Here we discuss just the direct and indirect pathways. Both originate with the striatum, which receives inputs from elsewhere in the brain and is organized so that nearby regions mostly receive inputs from adjacent brain areas. Striatal neurons are distinguished by whether they have D1 or D2 dopamine receptors (we will return to the significance of dopamine later). These give rise to the direct and indirect pathways, respectively. D1 neurons send inhibitory outputs directly to neurons in the output nuclei (hence, the name *direct*). The overall effect of D1 neurons is to decrease the inhibition generated by these output neurons, thereby activating the cortical and subcortical areas that they target. As an illustration, by stimulating appropriate D1 neurons, Roseberry et al. (2016) were able to activate neurons in the MLR that in turn generate locomotive activities.

The pathways starting with D2 neurons are referred to as *indirect*, as they involve intermediate nuclei. D2 neurons send inhibitory projections to the globus palladus external (GPe). Since GPe neurons also send out inhibitory projections, in this case to the subthalamic nucleus (STN), the net effect (inhibiting inhibition) is to increase activity of STN neurons. These STN neurons then send excitatory signals to neurons in the output nuclei, enhancing their inhibitory action. Overall, the effect of the indirect pathway is to strengthen the basal ganglia's inhibitory outputs. Accordingly, Roseberry et al. also activated D2 neurons, and demonstrated reduced MLR activity, which prevented specific forms of locomotion.

Due to their opposite effects of activating and inhibiting target areas, the direct and indirect pathways are thought to carry respectively Go and NoGo signals for specific actions (Hazy, Frank, & O'Reilly, 2007). Whether an action is selected depends on the relative activations of D1 and D2 neurons: roughly, if D1 activation (Go signal) is stronger, an action is selected, whereas if D2 (NoGo signal) is stronger, an action is inhibited. To perform this function, neither D1 nor D2 neurons need to encode any detailed information about specific actions; they only need to encode in the strength of the Go and NoGo signals the evaluation of the action generated elsewhere in the brain. What the basal ganglia do is effectively promote the action that receives the more positive evaluation and inhibits others (Bogacz & Gurney, 2007).

Among the brain areas affected by decisions made in the basal ganglia are areas in the neocortex. Indeed, neurons in the neocortex, basal ganglia, and thalamus are typically organized into loops in which output from the basal ganglia are directed back to the same locations from which inputs arise. The result is that the basal ganglia determine which among potentially competing processes in the neocortex are permitted to proceed (we will return to this role of the basal ganglia in determining the flow of processing in the neocortex in Section 9.3).

We noted earlier that the input neurons in the striatum are distinguished by the type of dopamine receptor that they contain. Dopamine, as we have discussed in Sections 2.3 and 4.2, is a volume transmitter that acts as a neuromodulator. That is, it alters the response properties of these neurons to their inputs. The importance of dopamine in the basal ganglia is illustrated by Parkinson's disease, in which low dopamine levels results in tremors and difficulty in initiating voluntary movement. Although claims about what modulatory role dopamine performs are still contested, we describe two of the more generally accepted hypotheses, one involving phasic (fast and temporary) and the other tonic (slow and continuous) dopamine signals.

Phasic dopamine signaling is widely viewed as enabling the basal ganglia to learn to make better decisions by representing reward prediction error – the difference between the predicted reward and actual reward following an action. Reinforcement learning is a type of learning that reinforces choices that result in rewards beyond those expected. The basic idea is that if the selected action generates the expected reward, no further learning is needed. If it generates more reward than expected (a positive reward prediction error), then that action should be reinforced. If it leads to less reward, then the selection should be attenuated. A phasic dopamine increase indicates a positive reward prediction error. It strengthens responsiveness of D1 neurons (the source of the direct pathway that carries a Go signal) and weakens the responsiveness of the D2 neurons (the source of the indirect pathway that carries NoGo signals). As a result, the action is more likely to be selected in the future. A phasic dopamine decrease results in the opposite effect.

In contrast, tonic dopamine levels are thought to control how an organism addresses the tradeoff between exploiting the current situation or exploring elsewhere for potentially better opportunities (Chakravarthy & Balasubramani, 2018). The example discussed in Section 2.3 illustrated this tradeoff in the context of seeking foods. The choice arises more generally: in selecting new music to listen to, a person must decide whether to continue with other compositions by the same artist (often successful when one enjoyed the current composition) or to explore those of other artists. One computational model suggests that lower tonic dopamine levels heighten the random fluctuation in the NoGo signals for the competing options, such that alternatives to the option represented as better will sometimes be selected, resulting in exploration. A higher level of dopamine dampens the fluctuation, resulting in the pursuit of the choice currently judged to be best (exploitation).

5.5 Summary

In this section, we have presented brief accounts of four areas in which neuroscience has provided an understanding of activity in the vertebrate brain. We will have occasion to return to each of these in subsequent sections.

6 How Do Neuroscientists Explain Activities of the Nervous System?

A major goal of any science is to explain phenomena – regularities in the world that researchers take to be important to account for (Woodward, 2019). Some phenomena are immediately obvious: organisms seek and eat food. But many phenomena are only discovered through extensive research. It took over 100 years of exploring electrical activity in animals to recognize that most neurons transmit action potentials (Section 2.1). Detailed studies using mazes were required for Tolman to conclude that rodents navigate using cognitive maps (Section 5.2). Sometimes researchers conclude that what were taken to be phenomena do not actually occur and so do not require explanation. For example, despite the claims of several researchers that worms could transfer what they learned to those that ate them, researchers concluded that this does not actually happen (for a philosophical analysis of this case, see Colaço, 2018).

In this section, we explore different accounts philosophers have offered of how scientists explain phenomena. Proponents of each account present them as characterizing explanations advanced in neuroscience. An important question is whether these accounts are competing or can be integrated into a common account. To the degree that they are competing, a further question arises: Does one need to choose between them or might there be multiple modes of explanation?

6.1 Mechanistic Explanation

Many biologists since the seventeenth century have viewed functional components within biological organisms as comparable to machines that humans construct. A key feature of machines is that they are composed of parts that carry out different activities. These activities are coordinated so that the whole machine generates a phenomenon that none of its parts alone can produce. Comparably, mechanistic explanations appeal to the composition and organization of a mechanism to explain a phenomenon (for further discussion of mechanistic explanations, see Machamer, Darden, & Craver, 2000; Bechtel & Abrahamsen, 2005; Craver & Tabery, 2019).

Much of the philosophical discussion of mechanistic explanation has focused on how researchers develop such explanations (Bechtel & Richardson, 1993/2010; Craver & Darden, 2013). The first step is to localize the phenomenon in a particular system – a mechanism – that is taken to be primarily responsible for producing it. For example, research on circadian rhythms in mammals (Section 5.1) identified the SCN as the responsible mechanism. Researchers studying spatial navigation in rodents identified the hippocampus as the locus of cognitive maps (Section 5.2). In these cases, deficits that resulted when the putative mechanism was damaged provided evidence for identifying the mechanism. In other cases, other strategies provided the basis for localization. Recording from neurons while presenting visual stimuli played a central role in determining the components of the mechanism of visual processing (Section 5.3), while the architecture of the basal ganglia motivated viewing it as a decision-making mechanism (Section 5.4).

After picking out the responsible mechanism, the key step in advancing a mechanistic explanation is to take it apart – to decompose it. There are two ways to do this – by identifying the mechanism's physical parts or by identifying the operations required to produce the phenomenon. Although identifying the physical parts is relatively straightforward with human-built machines, this can be challenging with biological mechanisms, as we saw in Section 2. Parts do not necessarily have well-defined boundaries. Golgi and Cajal debated whether neurons were separate units. Brodmann contested the traditional differentiation of cortical areas in terms of gyri and sulci, proposing instead that we use as boundaries the points where layers he identified in the neocortex changed thicknesses. One reason that identifying parts is challenging is that the goal is to pick out the parts that perform operations relevant to the phenomenon (what Craver, 2007, refers to as working parts). Not every way of cutting up the mechanism (e.g., chopping it into cubes) will be informative.

Differentiating operations requires different strategies from those used to identify parts. Consider, for example, the difference between distinguishing people and distinguishing occupations. One can distinguish people by their physical traits, but to distinguish occupations, one must determine what jobs the people perform. To identify the operations involved in generating a phenomenon, one needs both to reason about the operations that could produce the phenomenon and to find ways to intervene on the mechanism to see that they are indeed all carried out in the mechanism.

Reasoning from the phenomenon to the operations required is challenging. One of Gall's shortcomings that we did not emphasize in Section 3 is that he simply focused on traits on which people differ. He did not try to decompose these traits into the operations needed to realize them. Likewise, Broca characterized what the area that bears his name does in terms of the overall phenomenon – producing articulate speech – not the operations required to generate speech. The challenge in identifying operations is that, in most cases, operations are not described in the same vocabulary as the whole mechanism. Sometimes there is already a well-developed language for describing the operations performed by the parts of mechanisms (if individual neurons are the relevant parts, there is a rich vocabulary for describing their electrical and chemical activities). In many cases such a vocabulary does not exist and researchers must create it. This often requires proposing task analyses for the activity of the whole mechanism (if the mechanism performed operations A, B, and C then, by doing them in sequence, it would perform the overall activity) and then seek evidence that the mechanism actually performs those operations.

As difficult as it is to find one decomposition of a task, there are typically multiple task decompositions that would suffice to generate the phenomenon (this is why, once one designer creates a machine, competitors can often come up with their own designs that do not violate the original designer's patent). To figure out which is actually implemented in a given organism, researchers must try to localize the operations in different parts and then empirically investigate whether the parts actually perform those operations, using research strategies such as we described in Section 3.

In real science, there is often a prolonged period of revision both in the characterization of the parts and of the operations. For example, in developing the account of visual processing we described in Section 5.3, researchers subdivided Brodmann's areas 18 and 19 into V2, V3, V4, and MT/V5. And once researchers had advanced the hypothesis that MT/V5 is involved in motion detection, other researchers began to identify other operations it performs in addition to motion detection. In the end, though, the goal is to be able to map parts onto operations.

Beyond decomposing a mechanism into its parts and operations, there is a further step in characterizing the mechanism. Sometimes when we buy a product, the box will say "some assembly required." Until the parts are assembled, the product will not do what we bought it to do. Likewise, even if researchers had the correct account of the working parts of a mechanism and what operation each performs, they would not have a complete mechanistic explanation. Researchers must determine how they are organized so that the products produced in one operation can be further acted upon by other operations. To represent possible modes of organization, scientists often construct diagrams, such as Figure 14(b). In this diagram, rectangles represent brain regions while the icons in them represent the operations each is thought to perform. The lines connecting the rectangles, commonly referred to as *edges*, represent pathways between regions, with thickness indicating the prominence of the pathway. In other mechanism diagrams, yet other conventions are used (Abrahamsen, Sheredos, & Bechtel, 2018). While the diagram is static, it provides a basis for humans to reason about how the whole mechanism generates a phenomenon (in this case, recognizing an object or locating it in space) by performing the operations portrayed in the order shown. In this manner, researchers can mentally simulate the operation of the mechanism.

In thinking about the organization of a mechanism, humans start by thinking sequentially. The edges in Figure 14(b) are thought to carry activity from inputs at the bottom upward to higher processing areas. But the researchers who developed the diagram were very much aware that in the brain there are as many recurrent projections (neural projections from areas viewed as later in a pathway to those viewed as earlier) and that each of these areas sends and receives projections from regions of the thalamus and the basal ganglia (see Section 5.4). One can add additional rectangles and arrows to represent these, but it quickly becomes impossible to simulate the mechanism mentally. Instead, researchers often supplement a verbal and diagrammatic representation of a mechanism with a mathematical one, developing a computational model (Section 3.5). We will illustrate this in Section 6.3, but first we turn to an account of explanation that proposes using computational models to supplant the need for mechanistic accounts.

6.2 Dynamical Systems Explanations

Researchers in the life sciences often compare their sciences to physics. Explanations in many domains of physics appeal to laws that characterize how variables describing a system will change over time (hence, dynamical laws, often taking the form of differential equations). The explanation involves a demonstration that from the law and a specification of conditions at one time, one can derive what will happen at other times (Hempel, 1965). In many cases, the application of laws is far from simple and requires computational simulation to determine the consequences of the laws. Some cognitive and brain researchers apply similar strategies to explain behavior, and some philosophers have embraced these as fully legitimate explanations that do not require characterizing a mechanism.

A common approach of these investigators is to characterize a state space – a multidimensional space in which each dimension corresponds to a variable that describes the system. Consider three dimensions on which a gas can vary: pressure, volume, and temperature. Characterizing such a space would be of little explanatory interest if in fact the system could evolve from any point in the space to any other. What laws do is restrict the trajectory that the system can take through the space. The gas law:

$$\text{Temperature x Gas constant x Moles of gas} = \text{Volume x Pressure} \tag{6.1}$$

Imposes the restriction that when volume is held constant and temperature increases, so must pressure. If the actual system is shown to be similarly limited in its possible trajectories, then proponents of *nomological explanations* argue that the laws characterizing the possible trajectories through the state space explain why the system behaves as it does.

The gas law example (Eq. 6.1) does not specifically take time into account. But other laws spell out how values of variables will change over time. These give rise to what are termed *dynamical systems explanations*. Some dynamical laws, such as $x_{t+1} = x_t + 1$, are relatively simple: this law simply asserts that the value of the variable x increases by 1 at each timestep. If that is what happens, then the law explains why the variable follows this ascending trajectory. In many cases, the law will involve a more complex equation and produce surprising results. A mathematical function that is often employed to illustrate complex behavior is the logistic map function, $x_{t+1} = Ax_t(1 - x_t)$. The reader is invited to try various values of A between 3 and 4, picking an initial value of x_t between 0 and 1, and calculating the results for several steps. For example, with A = 3.3, values will initially fluctuate (a period referred to as the *transient*) but eventually begin to oscillate between two values (0.47943 and 0.82360). When A is increased to about 3.5, the values, after the transient, will jump sequentially between four values (approximately 0.49, 0.87, 0.38, and 0.83).[11]

These stable values are referred to as *attractors* – the idea is that values in their proximity will move closer to (fall into) the attractor. Figure 16(a) shows a two-dimensional state space in which there is just one fixed point attractor; initial values anywhere in the state space will fall into the attractor at the center. Sometimes attractors have a more complex structure, such as the cyclic attractor shown with a dashed line in Figure 16(b). In this case, no matter where the system starts, it arrives at a circle, around which it will progress indefinitely. Sometimes a space may have multiple attractors so that, starting from different points, the system may settle into different attractors. By representing a state space and identifying attractors in it, researchers can determine how the system will evolve from whatever point it currently occupies.

A much-cited dynamical model developed to explain animal behavior is the Haken–Kelso–Bunz (HKB) model of coordination dynamics. It describes

[11] The logistic map function is of interest because it can also demonstrate what is known as deterministic chaos – for most values above A = 3.6, the function will trace out a continually changing set of values without ever repeating, assuming one calculates the full real value of x. For illustrations, go to www.youtube.com/watch?v=ovJcsL7vyrk.

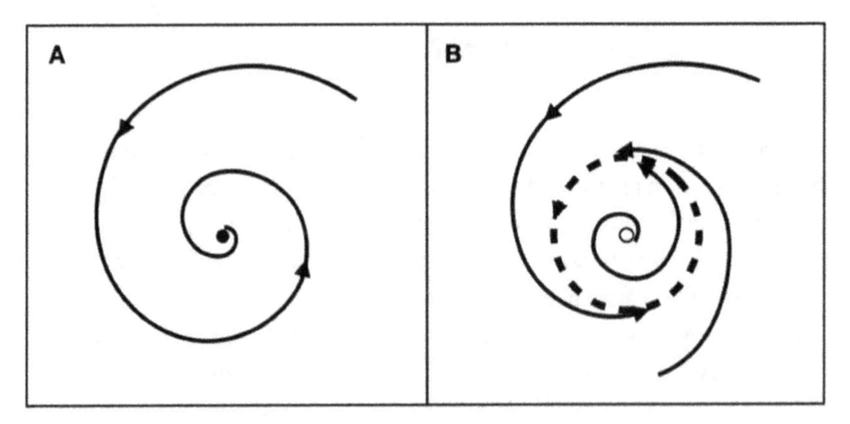

Figure 16 Attractors in a two-dimensional state space. (a) A point attractor. (b) A cyclic attractor.

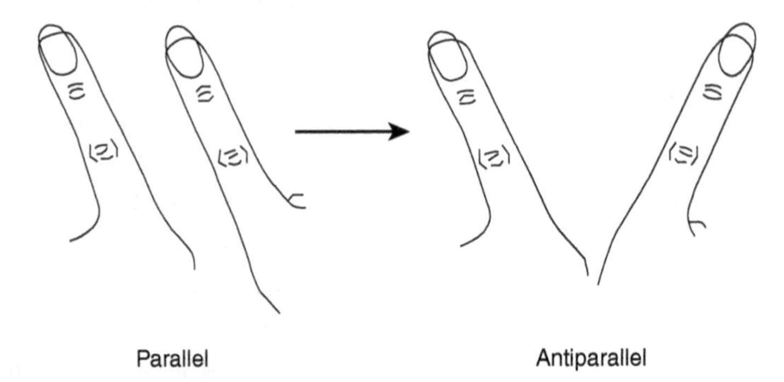

Figure 17 Parallel and antiparallel movement of fingers. Both can be maintained at slow speeds, but at faster speeds, only the antiparallel movement can be maintained. Figure by Hermann Haken released under the Creative Commons Attribution-ShareAlike 3.0 License.

phenomena such as the coordination between one's legs in walking (Haken, Kelso, & Bunz, 1985). To experience what the model describes, place both hands in front of you with the forefinger extended (Figure 17). Pretend your fingers are windshield wipers on a car. In most cars, wipers move in parallel (both tips move left, then both tips move right), but sometimes they move in an antiparallel fashion (the tips come together and then move apart). Try each pattern of movement, first slowly and then gradually faster. At slow-speeds most individuals can maintain both patterns, but when they try to speed up, they can only maintain the antiparallel pattern. The HKB model offers an explanation. It starts by describing the movement with the equation:

$$V(\phi) = -a \cos \varphi - b \cos 2 \phi \qquad (6.2)$$

in which ϕ is the phase relation between the fingers (or limbs more generally) and the ratio b/a is inversely related to the rate. In the state space described by Eq. 6.2, when b/a is high, corresponding to a slow speed, there are two attractors. However, when b/a is low, there is just one attractor. The loss of the attractor at faster speeds, on the dynamical systems account, explains your inability to maintain the parallel finger movement.

A notable feature of the HKB model is that the variable employed refers to a feature of the phenomenon (the angle between limbs), not to any proposed mechanism that is decomposed into components so as to account for the phenomenon. Proponents of dynamical systems accounts maintain that one does not need to enter into the nervous system to explain the inability to maintain the asymmetric movement. The phenomenon itself has structure that provides explanation (Chemero, 2000; Chemero & Silberstein, 2008). There are other examples where, merely from the characterization of the structure of phenomena, one can determine specific (and often unexpected) features of it. For example, from knowing the structure of tides around the ocean, one can determine that there must be a point in the ocean in which there is no tide. Likewise, from understanding how the circadian clock, discussed in Section 5.1, responds to light stimuli, one can infer that in some organisms, there is a time at which exposure to light will cause the amplitude of the oscillations to become 0 (which is to say, the clock will stop, as there is no longer an oscillation to represent time). This necessity was in fact demonstrated before the mechanism of circadian oscillation was known and does not depend on any details about the mechanism (Winfree, 1987; for discusssion, see Bechtel, 2021).

6.3 Dynamic Mechanistic Explanations

As we noted in discussing mechanistic explanations in Section 6.1, when mechanisms depart from sequential organization, it becomes challenging to simulate their behavior mentally. Even a simple feedback loop can present a challenge. As many people are aware from examples like thermostat controlling furnaces, feedback loops can generate oscillations (the temperature will rise after the thermostat turns the furnace on and fall after it turns it off). Accordingly, when an intracellular feedback mechanism was proposed to explain circadian rhythms (Section 5.1), it was expected to generate oscillations. But the question was whether the oscillations would be sustained indefinitely or dampen over time. To address that question, Goldbeter (1995) created a computational model that showed that under biologically plausible conditions, the mechanism would instantiate a cyclic attractor (Figure 16(b)). He therefore concluded that the biological mechanism implementing feedback could oscillate indefinitely. This explanation is much like that provided by the HBK model in Section 6.2, but here the variables refer to hypothesized operations of the components of the mechanism. Since in this case the explanation is a hybrid, drawing upon both mechanistic decompositions and the use of computational models to characterize the dynamical behavior of the mechanism, Bechtel and Abrahamsen (2010) refer to them as *dynamic mechanistic explanations.*

Perhaps the best-known dynamical computational model in neuroscience is the Hodgkin–Huxley model of the action potential. Hodgkin and Huxley (1952) decomposed the current across the neuron membrane into components for sodium, potassium, and other ions and developed an equation for how each contributed to the current *I* across the membrane (Figure 18). From this they produced an overall equation that described how the current changes as the electrical potential changes:

$$I_m = C_m \frac{dV_m}{dt} + \overline{g_K} n^4 (V_m - V_K) + \overline{g_{Na}} m^3 h (V_m - V_{Na}) + g_l (V_m - V_l). \quad (6.3)$$

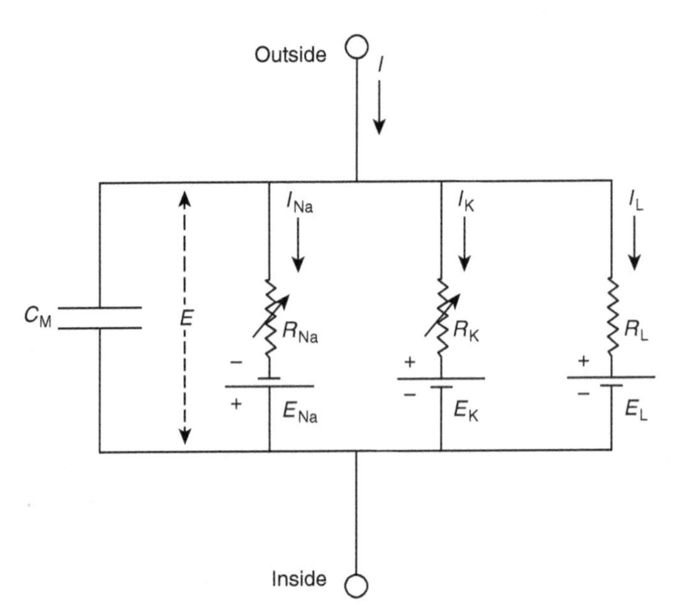

Figure 18 Hodgkin and Huxley's (1952) representation of the current *I* across the membrane in terms of the currents for sodium (Na), potassium (K), and leakage (l) due to other ions. *E* represents the membrane potential and *R* the resistance for each ion.

In Eq. 6.3, I is the current, C_m is the capacitance due to the membrane, V_m the electrical potential across the membrane, V_K, V_{Na}, and V_l represent the potential due to potassium, sodium, and leakage (other ions), and g_K, g_{Na}, and g_l the conductance for the various ions. n, m, and h are parameters used to fit the model to data. From Eq. 6.3, one can generate the pattern of the action potential (Figure 2).

Hodgkin and Huxley's accomplishment, which earned them the Nobel Prize for Physiology and Medicine, has been the focus of considerable philosophical controversy. Weber (2005) treated it as an instance of an explanation that derives a phenomenon from a law. In the nomological tradition, laws are typically distinguished from causal claims, and Weber (2008) subsequently offered a revised account, according to which Hodgkin and Huxley offered a causal explanation in which the ion currents caused the action potential. While granting the usefulness of the model as a description of the action potential, Craver (2006, 2008) has argued that it is not explanatory since it does not include, let alone characterize, what he takes to be the critical parts of the mechanism generating the action potential, the gates on the channels through which ions are allowed to enter or leave the neuron. It turns out that the coefficients of the parameters n, m, and h correspond to features of these gates, but this was only discovered years later. At best, Craver allows, Hodgkin and Huxley offered a sketch of an explanation that was only provided later. More recently, Levy (2013) has argued that the model does in fact provide a mechanistic explanation in so far as it presents the whole current as arising from the aggregate activity of each of the types of ions. The second, third, and fourth summed terms in the equation represent the current generated by each ion as a result of the difference between its current potential and the membrane potential. Levy contends that the Hodgkin–Huxley model captures the crucial activities in the mechanism. As a result, it offers a dynamical mechanistic explanation of how the changing concentrations of the ions give rise to an action potential. This debate illustrates different stances philosophers take on the nature of explanation and what is required to explain a phenomenon.

6.4 Network and Connectomic Explanations

As we have seen in various sections of this Element, the nervous system, and its various subparts, are often characterized as networks. The crucial idea of a network is that it consists of entities (represented as nodes) and connections between them (represented as edges). Networks are ubiquitous – any time entities are connected, they can be represented as a network. But some networks

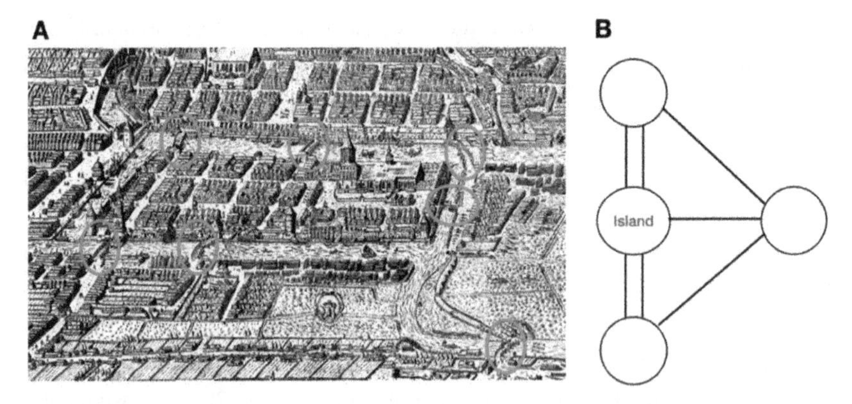

Figure 19 (a) Map of Konigsberg with the river and seven bridges highlighted by Bogdan Giuşcă and distributed under the Creative Commons Attribution-Share Alike 3.0 Unported license. (b) Network graph, in which nodes represent different landmasses and edges the bridges between them.

have distinctive properties that are sometimes viewed as explaining aspects of the behavior of the system instantiating the network.

One of the earliest examples of a network analysis was Leonhard Euler's solution to a problem posed by the bridges crossing the Pregel river in the Prussian town of Konigsberg (Figure 19(a)): Can one cross each bridge just once on a walk? He represented the different landmasses with a node and the bridges with an edge (Figure 19(b)). From this abstract representation, Euler proved no route is possible. For it to be possible to cross each bridge just once, each node other than the ones representing the starting and ending locations must connect to an even number of bridges. In this case, all four nodes connect to an odd number of bridges; accordingly, such a walk is not possible.

Starting in the mid–twentieth century, investigators identified a number of important features of networks that determine the properties of any actual system instantiating the network. Here we introduce just two concepts that turn out to be extremely important for understanding the brain: small worlds and hubs. To introduce these, we need to introduce some of the measures used to describe networks. One is the average of the shortest paths between each two nodes. A second is how clustered a network is: to how many of its neighbors a node is connected. In a randomly connected network, the average shortest path is short but clustering is low. In a regular lattice (a structure in which every node is connected to each of its neighbors), clustering is high (since a node is connected to all its neighbors) but the average shortest path is long. Watts and Strogratz (1998) showed that many networks in the real world are more like the

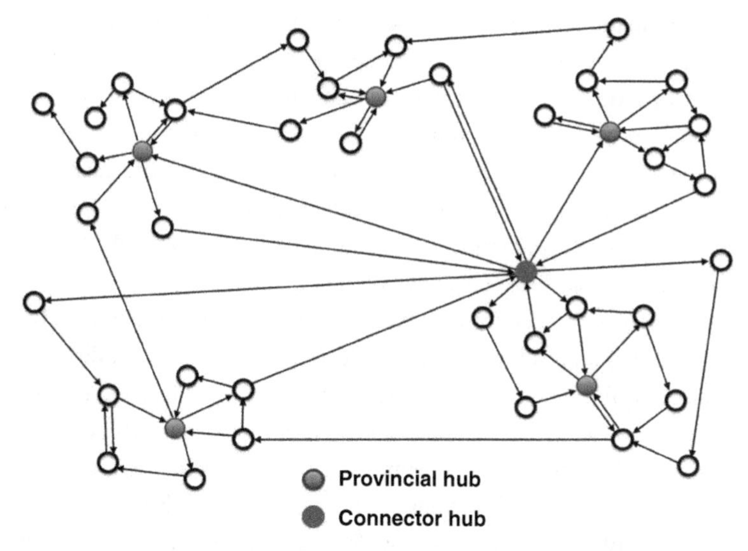

Figure 20 A network with relatively short average path between any two nodes, relatively high clustering, with some nodes having many more connections than others and hence serving as hubs.

one in Figure 20, in which the average shortest path is relatively short but nodes are also highly connected to their neighbors (collections of nodes that are highly connected to each other are often called *modules*). They call these networks *small world networks*.

A third measure is how the *degree*, that is, the number of connections from each node, is distributed. If degree is distributed normally (i.e., if values are equally distributed about the mean and decrease with distance from the mean), no node will be especially highly connected. But in many real world networks, Barabási and Bonabeau (2003) showed the degree is not distributed normally but according to a power law (a mathematical relation of the form $y = ax^{-k}$). This results in a few nodes being highly connected while most have few connections. As illustrated in Figure 20, those highly connected nodes can be the basis for a local module (provincial hubs) or can serve to integrate modules (connector hubs).

A number of researchers have analyzed nervous systems in network terms. In Section 4.2, we described how researchers produced a complete connectome for the nematode worm *C. elegans*. This network turns out to have small world properties. Developing connectome representations for other species at the level of individual neurons is extraordinarily challenging, although researchers are getting very close to having such a map for the fruit fly (which has about 100,000 neurons). Instead, researchers concerned with connectivity in the neocortex of mammals have focused on connections between brain areas (e.g., BAs) and are analyzing these for their properties (Sporns, 2010, 2012). The principles of short average path length, high clustering, and hubs all appear to apply. Van den Heuvel and Sporns (2011) have further shown that the human brain instantiates a rich club structure – a set of regions, each of which serves as a hub, are more connected than would be expected even given their high degree. Given their network properties, these brain regions are thought to serve as a communication backbone for the whole brain.

Given the potency of concepts such as these to explain activity in networks, Huneman (2010) argues for treating topological explanation as a distinct form of explanation. In particular, he distinguishes it from mechanistic explanation since it does not focus on the contribution of parts but only on how they are connected. In more recent work, Huneman (2018) has addressed how topological and mechanistic explanations can be integrated. The basis for integrating them is that topological principles provide a basis for understanding the consequences of different modes of organization in biological mechanisms. When topological principles such as small world organization suffice to account for the phenomenon, it is the organization, not features of the individual components, that explain the behavior of the mechanism (Levy & Bechtel, 2013).

6.5 Control Mechanistic Explanations

In philosophical discussions, mechanisms are often portrayed as ready to operate whenever their start or setup conditions are realized (Machamer, Darden, & Craver, 2000). To experiment on mechanisms using techniques such as those introduced in Section 3, researchers try to set up conditions in which they do operate in a regular manner. However, in an organism, the continuous operation of a mechanism is often not needed and can in fact be harmful (just consider continually contracting the muscles in your legs). Instead, mechanisms need to be activated and deactivated as needed by the organism. The same is true of the machines human make. We do not desire a furnace to produce heat all the time. Accordingly, we employ thermostats that turn the furnace on when the temperature drops too low and off when it is warm enough. The thermostat represents a second machine that operates on the primary one, changing some of its parts so that it operates in different ways at different times. Biological organisms are replete with mechanisms that operate on other mechanisms. That is, in fact, what neurons and neural mechanisms do: they control the operation of other mechanisms such as muscles and glands.

There is an important difference between biological mechanisms and human-built machines. We design machines to be controlled by us. We turn our car engine on or off, and when on, we control the fuel supplied to it through depressing the accelerator. Who controls biological mechanisms? The short answer is the organism itself. Recognizing this, Maturana and Varela (1980) introduced the crucial idea that organisms are autopoietic: they build themselves by procuring matter and energy from their environments and directing it into the synthesis of their own bodies. This requires control over procurement and construction mechanisms. In addition, the tissues that make up organisms are prone to break down, requiring organisms to detect failures and deploy repair mechanisms (Rosen, 1972). In virtue of constructing and repairing themselves, organisms are sometimes referred to as *autonomous systems* (Moreno & Mossio, 2015).

Organisms are not agents over and above the mechanisms that constitute them. Autonomy results from the actions of the mechanisms constituting an organism. More specifically, it results from the deployment of control mechanisms. Like a thermostat, control mechanisms act on and change the configuration of other mechanisms in light of conditions either in the organism or its environment (Winning & Bechtel, 2018). To do this, control mechanisms must make measurements (or utilize measurements made by other control mechanisms upstream of them). The measurement component of the control mechanism results in the state of the control mechanism being determined by the value of the variable being measured. Again, the thermostat provides a model – a component internal to the thermostat is altered by the temperature in the environment. Given the measurement, the control mechanism produces a specific action on the controlled mechanism. This means that control mechanisms must be properly configured so that the changes that they make in other mechanisms are appropriate to the circumstances that the organism faces.

The word *autonomy* includes the Greek words for self (*autos*) and law (*nomos*), and thus signifies that an autonomous system sets laws for itself. Civil laws set norms for behavior. In determining the behavior of other mechanisms, control mechanism likewise impose laws or norms that govern that behavior (Winning, 2020). In the case of the thermostat, these norms ultimately derive from the humans who build and set the thermostat. Biological control mechanisms are not designed by humans. Rather, they are the product of evolution. In the course of evolution, those control mechanisms are retained that apply norms that enable organisms to maintain themselves and reproduce. Those that do not disappear over the course of evolution.

One reason control mechanisms are so crucial to living organisms is that organisms regularly confront different circumstances that require different responses. They need to be able to adapt to these. Some circumstances repeat and, like a thermostat, control mechanisms can direct the same response on each occasion. But organisms often confront novel situations that require tailoring their basic mechanisms in new ways. To deal with these situations, control mechanisms must exhibit a degree of flexibility, directing basic mechanisms to operate in novel ways. In Section 10, we will explore ways in which control mechanisms are organized so as to support creating effective responses to novel situations.

6.6 Summary

We have introduced several different perspectives on explanation: mechanistic, dynamic, and topological. Each appeals to different factors and seems applicable to specific phenomena. This suggests a pluralistic perspective, recognizing different types of explanation. It also suggests that different perspectives might be integrated, and we offered dynamic mechanistic explanations as one integrated perspective. Lastly, we noted the importance of control in biological organisms and described how the mechanistic perspective can be extended to characterize control mechanisms.

7 What Are Levels in Neuroscience and Are They Reducible?

The term *level* is widely invoked in neuroscience, and researchers and commentators often debate whether some levels should be reduced to others. Unfortunately, the term level is used in a wide variety of senses. In this section, we differentiate three notions of level that are prominent in discussions about neuroscience and identify the implications of each for reduction.

7.1 Marr's Levels (Perspectives)

David Marr, a pioneer in the development of computational modeling in neuroscience (Section 3.5), began his book *Vision* (1982) with a critical assessment of what he saw as the current state of the discipline. Neuroscientists were accumulating many findings about how various parts of the nervous system operate using techniques such as those discussed in Section 3. But they were making little progress in providing an understanding of how the brain works. On his analysis, this was due to focusing on just one level, which he termed the *hardware implementation* level. Accounts at this level focus on parts of the brain and how each operates. To make progress in understanding the brain, he argued for the need for two other levels: those of *representation and algorithm* and of *computational theory*. At the representation and algorithm level, he argued that researchers should treat the parts of the brain as representing content and applying rules to manipulate those representations. Much of Marr's own work was focused on the representation and algorithm level as he attempted to describe visual processing in terms of states that represent specific features of stimuli and algorithms that specify how the brain operates on one representation to generate another (see Section 8). But the most novel, and arguably the most important, of his levels was that of computational theory. At this level he proposed that researchers should address questions such as "What is the goal of the computation, why is it appropriate, and what is the logic of the strategy by which it can be carried out?" (p. 25).

To address the goal and appropriateness of neural computation, it is not sufficient to look inside the nervous system any more than one can determine the goal and appropriateness of what goes on in a computer by just looking inside it. With the computer, we turn to the users and the tasks for which the computer is used. The comparable move with respect to the nervous system is to look to the environment in which it works – the organism and the physical and social environment in which the organism operates. As Marr's focus was on vision, the relevant environment is the visual world. His contention is that by analyzing the structure of the visual world, we can understand what the visual system needs to do. Marr notes that James Gibson, a psychologist with whom he mostly disagreed, came closest to understanding what it meant to focus on the computational level. Gibson (1979) argued that visual experience does not consist of independent pixels but is highly structured. An illuminating example is that if an object is approaching you, or you it, it expands in your visual field. If a ball is expanding equally in all directions, then you may intersect it. Whether you do and how soon depends on how fast it expands. Studying the challenges an organism faces in its environment, Marr insisted, is critically important to understanding what a nervous system is doing and whether what it does is appropriate (Shagrir, 2010).

Marr's levels might better be glossed as perspectives that investigators need to take in understanding the nervous system. One perspective is to focus on parts of the nervous system and how each is operating. A second focuses on the procedures that the brain is using to process sensory information or generate actions. The third focuses on the organism and its environment in order to identify the tasks that the nervous system must perform if the organism is to be successful. Each perspective may inform one of the others (knowing the task that the nervous system must perform can guide the search for representations and algorithms that operate over them). However, one perspective cannot provide the insights provided by another. Rather than reducing one level or perspective to another, Marr argues that researchers need to pursue all three.

7.2 Mechanistic Levels and Reduction

The framework of mechanistic explanation (Section 6.1) brings a different conception of levels: the components of a mechanism can be viewed at a lower level than the mechanism itself. Likewise, mechanisms can be constituents of mechanisms at higher levels. This look down to lower levels and up to higher levels iterates due to the fact that mechanisms are often parts of yet higher-level mechanisms and their parts are themselves mechanisms consisting of parts. The notion of part and whole is fundamental to this notion of level. As a result of the process of decomposing mechanisms into other mechanisms, mechanistic levels are hierarchical.

In decomposing mechanisms, researchers are going down levels. How far should they go? There is a long tradition in science, referred to as *reductionism*, which argues that explanation should appeal to as low a level as possible because lower levels are, in some sense, more basic. For some theorists, the most basic level is that of fundamental physics; for these theorists, the ultimate goal is to explain all happenings in the universe, including those in our brains, in terms of the entities and activities of the most basic physical particles. For now, there seems to be little prospect of explaining biological phenomena in such terms. Nonetheless, some theorists argue that the goal should be to explain behavior and cognition at the lowest level possible. Bickle (2006) defends what he terms *ruthless reduction*: intervene at the lowest level possible (currently the molecular level) and measure effects at the behavioral level. When successful, this provides evidence that entities at the molecular level are causally efficacious and so explain the behavioral phenomenon.

Mechanistic explanations are clearly reductionist in one sense: they appeal to the components of a mechanism in explaining its behavior. And insofar as the components of a mechanism are themselves mechanisms, they support going to yet lower levels. This is conveyed in Figure 21: to explain the behavior of a mouse navigating the Morris Water Maze (Section 5.2), neuroscientists appeal to the hippocampus as the locus of spatial maps, to synapses between neurons in which chemical changes are realized (long-term potentiation), and finally to NMDA (N-methyl-D-aspartate) receptors in the postsynaptic membrane. Yet, it is important to realize that at each lower level, the component is one part of a mechanism: A single synapse does not realize a cognitive map, but only in the context of other neurons in the hippocampus. Indeed the hippocampus does not operate on its own but only in the context of a larger mechanism involving other brain areas such as the entorhinal cortex (Bechtel, 2009). At each of these levels, components are organized into larger mechanisms that do things that the components cannot do. While appealing to lower levels, mechanistic accounts

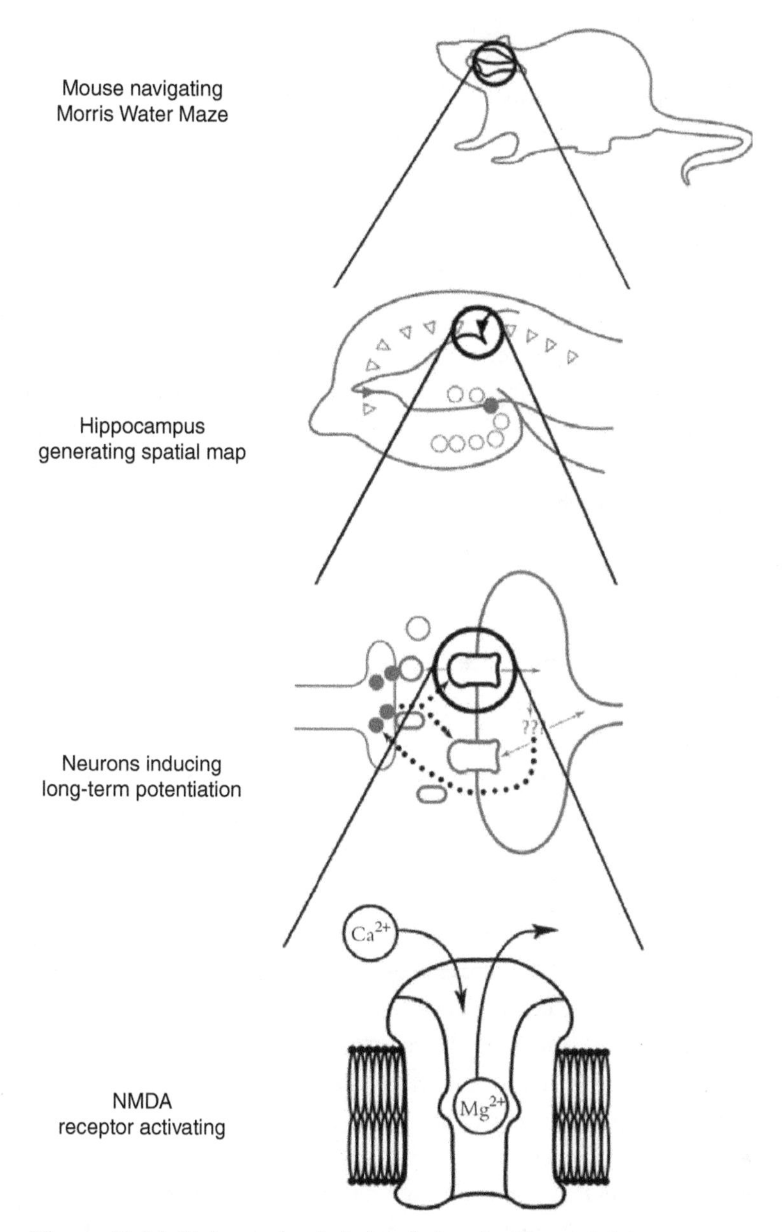

Figure 21 Multiple mechanistic levels invoked in explaining memory. Reprinted from Craver (2007) with permission of Oxford University Press.

do not privilege any one of them. Rather, mechanistic accounts emphasize equally how components are integrated into larger systems. One reason it is important to go up levels is that in many cases, lower-level components perform different operations when they are part of different higher-level wholes.

7.3 Levels of Control

One of the factors that can make a lower-level mechanism behave differently is that control mechanisms (Section 6.5) operate on it. Control mechanisms give rise to their own relation of levels. Insofar as a control mechanism operates on and changes the parts or operations of another mechanism, one can view it as at a higher level. And insofar as another control mechanism operates on it, that control mechanism is at a yet higher level. But this relation differs from the relation between mechanistic levels in two respects. The control mechanism is not a whole containing the controlled mechanism. And although the relation between control mechanisms can be hierarchical, it need not be (see Section 10.2). Whether hierarchical or not, levels of control do not give rise to reduction, as each control mechanism has its own role to play in coordinating the activity of the mechanisms it controls.

7.4 Summary

We have identified three notions of level that figure in discussions of neuroscience. Only the mechanistic conception gives rise to a notion of reduction. Some theorists advocate advancing explanations at the lowest level possible. Most proponents of mechanist explanation, however, emphasize the importance of organization at each level, and so recognize contributions of both lower and higher levels.

8 Do Neural Processes Represent Anything?

Representation is perhaps the most contested term in philosophical discussions of neuroscience. A representation stands in for and can be used by a process as a surrogate for something it represents. Humans frequently make use of representations. The phrase "nervous system" represents nervous systems and is used in many sentences in this Element. Many of the figures in this Element represent parts of the nervous system while others represent processes thought to occur in it.

On many accounts, such as the one advanced by Marr (Section 7.1), the nervous system is a computational system. A key component of a computational account is that operations are performed on representations (philosophers often refer to this as the *computational theory of mind* – see Pitt, 2020). Accounts of neural activity commonly characterize that activity in terms of what it is supposed to represent. Thus, place cells (Section 5.2) are characterized as representing places – when they spike in sequence, they indicate a sequence of locations on an animal's route. Neurons in different regions of the visual cortex (Section 5.3) are characterized as representing edges, shapes, motion, and so on.

Neuroscientists certainly ascribe representations to neural processes. What is contentious is whether neural processes actually do represent objects and events. If so, what do they represent?

8.1 Can One Account for Neural Processes without Invoking Representations?

In Section 6.2, we introduced the dynamical systems approach to explanation that eschewed the need to appeal to mechanisms to explain neural processing. In eschewing mechanisms, this approach also rejects the idea that the brain should be viewed as a computational system that operates over representations. As an exemplar of an alternative to a computational system, van Gelder (1998) cites the governor invented by James Watt to control a steam engine so that it would maintain a constant speed regardless of what appliances were attached to it. In the governor, weights are attached by flexible spindle arms to a spindle that rotates with the flywheel of the steam engine (Figure 22(a)).[12] The arms move up or down, depending on the centrifugal force generated by the turning spindle. As a result of the linkage mechanisms connecting the spindle arms to the steam valve, the activity of the arms slows the steam flow when the flywheel is turning too fast (due to relatively strong steam flow and weak resistance) and increases it when it is turning too slowly. The governor thus uses negative feedback to control the steam engine.

To make his argument that representations are not needed to explain the Watt governor, van Gelder appeals to the mathematical characterization developed by Maxwell (1868):

$$\frac{d^2\varphi}{dt^2} = (n\omega)^2 \cos\varphi \sin\varphi - \frac{g}{l} \sin\varphi - r\frac{d\varphi}{dt} \tag{8.1}$$

in which ϕ is the angle of the arms, n is a gearing constant, ω is the speed of the engine, g is a gravity constant, l is the length of the arms, and r is the friction of the hinges. Equation 8.1 does describe the behavior of the governor and, on a dynamical systems account of explanation (Section 6.2), it provides an explanation. Importantly, there is no reference to representations.

A counterargument that one does need representations to understand Watt's governor starts by noting that the dynamical account does not answer the question of why Watt inserted the spindle arms into the governor (other than

[12] For a video illustration of the operation of the Watt governor, see www.youtube.com/watch?v=B01LgS8S5C8.

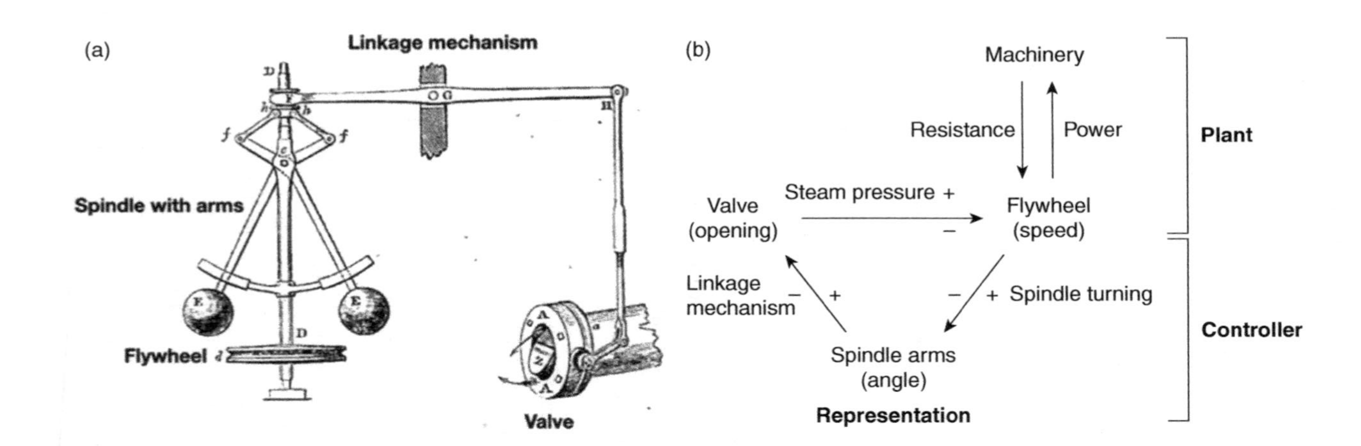

Figure 22 (a) Watt's design of the Watt governor. (b) Schematic representation showing how spindle arms represent the speed of the machinery and how this representation is used to affect the valve opening.

by saying that this allowed him to create a system that satisfies Eq. 8.1). A mechanistic explanation, by focusing on what the parts of the governor do, offers an explanation by appealing to the fact that the spindle arms stand in for the changing speed of the engine in a mechanism that can use them to adjust the speed of the engine (by acting on the valve). Precisely what they represent is not straightforward, but Nielsen (2010) showed that by solving Eq. 8.1 for ω, one can show that the speed is represented by $\varphi, \frac{d\varphi}{dt}$, and $\frac{d^2\varphi}{dt^2}$, that is, in terms of the angle of the spindle arms, the velocity with which it is changing, and its acceleration.

In addition, Watt's governor fits the characterization of a control mechanism. As discussed in Section 6.5, control mechanisms alter the operation of other mechanisms, in this case, by opening or closing the valve on the steam pipe, based on the measurements that they make (Figure 22(b)). Seen from the control mechanism perspective, Watt's genius was to realize that he could take advantage of centrifugal force so that the angle of the spindle arms would measure the speed of the engine. As a result of performing this measurement, the angle of the arms stands in for and so represents the speed of the engine and does so in a format that enables the governor to alter the valve setting so as to maintain the target speed. More generally, the fact that control mechanisms make measurements in order to perform their control function is what leads researchers to characterize them as employing representations – they represent what they measure. Neural mechanisms that respond to fat concentrations in the intestinal system (Section 5.1) or moving objects in one's visual field (Section 5.3) make such measurements, and so, on these accounts, can be characterized as representing these conditions.

8.2 Are Representational Ascriptions Mere Glosses by Theorists?

One might acknowledge that neuroscientists attribute representations to components of control mechanisms but insist that these are only glosses provided by theorists. Representations do not figure in the operation of the governor or the brain: the physical processes in the governor are not being used as representations by these physical systems (Haselager, de Groot, & van Rappard, 2003; Egan, 2019). The governor would work the same if disconnected from the flywheel and the steam valve as long as something turned the spindle arm. It could even be used to control some other process. One way to put this challenge is that the angle arms do not know about the flywheel or that it is acting on a steam engine, and so do not really represent it. (For a classic argument of this type directed not at brain mechanisms but artifical intelligence systems, see Searle, 1980.)

In introducing representations, we used linguistic phrases as examples. How does a word like "neuron" represent neurons? There is nothing about the word itself that determines what it represents. Waldeyer, who invented this term, might have coined a different term. Rather, its meaning depends on the conditions in which language users insert it into sentences and, especially, how they respond to it when they encounter it being used by others. This raises the question as to whether a similar account applies to neural activities characterized as constituting representations. Viewing neural activities as states in control mechanisms already suggests how a similar account can be developed: When they are generated and used in controlling other mechanisms, neural mechanisms treat them in ways similar to how language users treat linguistic phrases.

Further support for interpreting neural processes as representations is that working scientists investigate how they possess the ability to represent. When you encounter a word or phrase you do not know, you investigate when people use it and how others respond to it. Similarly, after O'Keefe characterized neurons in regions of the hippocampus as place cells, he and others set out to identify how they come to respond to particular locations and how they figure in the animal's behavior. As discussed in Section 5.2, researchers manipulated environments to see which would produce changes in a neuron's response. They also showed how activation of these neurons before and after rodents ran on paths occurred in anticipation and recall. At a minimum, the researchers are not just glossing an already described neural mechanism, but taking seriously the hypothesis that it functions as a map-like representation and investigating how it does so. That is, they, treat these neurons as actually representing places (Bechtel, 2016).

8.3 Do Neural Processes Accurately Represent the World?

If one accepts that neural processes do represent the world, a further question is: Do they *accurately* represent the world? Many of the approaches to ascribing content to neural processes assume that that is what they are doing. In characterizing what different visual areas do, researchers presented stimuli with a given feature and treated the neurons that responded most strongly as representing that feature. One might also wonder what would be the point of a perceptual system that did not accurately represent the world since that would seem to defeat the goal of successful interaction with the world.

Using temperature perception as an example, Akins (1996) argues that our sensory systems do not accurately represent the world. Rather, she characterizes our temperature receptors as narcissistic – they respond when the temperature of objects we are touching is too hot or too cold. Moreover, they do so in a nonlinear fashion: as temperature increases, the hot receptor will first generate spikes at a greatly increased frequency before gradually dropping to a frequency slightly above what it was at the outset. When the stimulus terminates, it stops spiking altogether before gradually increasing to its default rate. The cold receptor operates similarly. Such a system signals major changes in temperature, but unlike a thermometer, it does not give a readout that corresponds to the temperature of the object touched. Moreover, responses are contextually sensitive to immediately preceding experiences, as illustrated in a familiar demonstration: put one hand in hot water, the other in cold, and then move both to water of intermediate temperature. It will feel like the hand previously in cold water is in warmer water than the one originally in hot water.

While such a sensory system does not generate accurate representations of temperature, Akins argues that it does alert the organism to what it needs to know – is something too hot, such that I should drop it or avoid touching it? Receptors that acted like a thermometer would be less efficient – the nervous system would have to incorporate the temperature information into a plan for action rather than responding directly. Does Akins' argument support treating the senses and the neural processing systems downstream from them as generating narcissistic representations or alternatively, as "nonrepresentational systems"? In fact, Akins has argued for the stronger, nonrepresentationalist, conclusion: sensory systems are not only narcissistic but function through nonrepresentational feedback processes. However, it is not clear why one cannot view the activity of the sensors as constituting context-sensitive narcissistic representations. Consider again the Watt governor – one may argue that it does not represent the speed of the flywheel per se, but only represents, narcissistically, that it is moving too quickly or too slowly.

8.4 Summary

We have characterized different views about whether brain processes count as representations. What is worth highlighting is how these views tie to views of explanation discussed in Section 6. If one adopts the view that the nervous system embodies control mechanisms, one needs to characterize it as making measurements, and hence talk of representations seems motivated. If one adopts a dynamical system view that eschews mechanisms, then one can view these processes as elements of dynamical systems without treating them as representations.

9 What Is Distinctive about the Neocortex?

As we noted at the outset, the neocortex is the brain area that has expanded the most in primates, including humans. It is clearly important for human life, especially for those activities that humans distinctively perform. But, as we have stressed through this Element, other brain structures are also important. With few exceptions, the neocortex does not take over their activities but supplements them. The relevant question is: What is the distinctive type of processing that occurs in the neocortex? One suggestion comes from the studies of decorticate cats discussed in Section 5.4. While cats in which the neocortex is removed can live in protected environments, they would be unlikely to fare well in the world in which they confront variable conditions, including predators. Based on these studies, Buchwald and Brown (1973) proposed that the neocortex serves for detailed analysis of stimuli, extracting and representing complex and subtle information about an organism's environment and identifying relations between different bits of information. Such information is extremely useful in solving problems posed by a variable environment. In this section, we investigate how the neocortex can perform these tasks.

9.1 (Artificial) Neural Networks and Pattern Extraction

The neocortex is organized in a distinctive manner that supports the hypothesis that it acts to extract subtle and complex information from sensory inputs. While many brain areas, including both the basal ganglia and the hypothalamus, are organized as interconnected nuclei, the neocortex is laid out much more systematically. As we discussed in Section 2.4, Brodmann (1909/1994) differentiated areas within the neocortex based on the thickness of layers identified in stained cortical tissue. Tracing axons from neurons in one area reveals that they mostly project to selected neurons in other specific areas, resulting in relatively orderly anatomical hierarchies such as shown in Figure 14(b). At the top of the figure are areas in the temporal and parietal lobe. Both streams, however, continue into the frontal cortex, reaching the far frontal area known as the *prefrontal cortex*, on which we will focus in Section 9.4.

To see how such a (anatomically) hierarchically organized network could enable the extraction of information, consider *artificial neural networks* (ANNs) – computational systems that were inspired by the architecture of the neocortex. As illustrated in Figure 23, these networks consist of layers of artificial neurons, commonly referred to as *units*. A weighted connection links a unit in one layer to units in the next higher layer; in processing, the weight is multiplied by the activity value of the unit in the lower layer to determine an input to the higher-level unit. Each higher-level unit accumulates these inputs

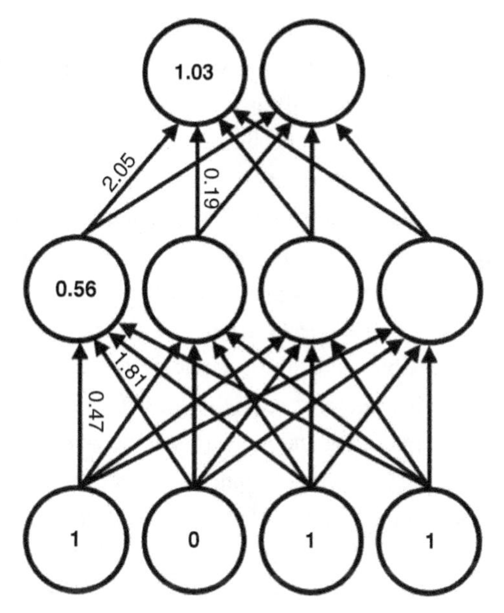

Figure 23 Simple artificial neural network. Activation values for the input units at the bottom are multiplied by weights on the connections (indicated by arrows) to determine the activation of units in higher layers. Example weights and activation values are shown.

and applies a nonlinear mathematical operation to determine its activity value. Such a network will generate output activity from values supplied on its input layer and can be trained to generate desired outputs for different inputs. A common way to train ANNs (known as *backpropagation*) is to let the network generate an output from whatever weights it has and then to apply an algorithm to gradually change weights so as to reduce the difference between the actual and desired output. Over multiple iterations of training, such networks can learn to respond similarly to different instances of the patterns. For example, a network can learn to recognize pictures containing different species of dogs. As a result, they are often characterized as *recognizing patterns*. When successful, these networks can generalize and recognize patterns when tested with novel stimuli (Bechtel & Abrahamsen, 2002; Buckner & Garson, 2019). In recent years, researchers have developed *deep-learning networks* that employ numerous layers of units between the input and output, with the weights on different layers of connections each able to be adjusted during learning to achieve better performance (Sejnowski, 2018). An intriguing finding is that when deep-learning networks have been deployed to model processing of visual stimuli, they end up acquiring an organization of nodes that is similar to that found in the human visual system (Yamins & DiCarlo, 2016).

Artificial neural networks are powerful systems for recognizing patterns. Since the task of vision is to extract patterns in visual stimuli, it is not surprising that perceptual processing areas in the brain are organized in the same manner. It is also easy to see how an ANN-style architecture can implement motor control: allow inputs to encode a high-level description of an action and the network can be trained to generate outputs that implement the specific motor activities required. One can extend pattern recognition beyond perceptual and motor processing to more clearly cognitive tasks that involve a sequence of inferences. Each layer in a network can be viewed as making an inference based on inputs from the previous layer (e.g., infer that an object is a bird), providing the next layer an input from which it can make a further inference (e.g., that it can fly). Accordingly, ANNs are widely used to perform reasoning and problem-solving tasks, and the fact that the neocortex is organized in a sequence of connected layers suggests that it carries out reasoning and problem-solving activities in a similar manner.

9.2 The Challenge of Explaining the Systematicity of Thinking

When theorists advanced ANNs as models of human cognitive processing in the 1980s, they were confronted by a host of objections, one of which is that ANNs cannot account for what is termed the *systematicity* of human thought. Systematicity is exemplified in arguments used to establish conclusions in such fields as mathematics, law, and science. Consider the argument:

(1) A dog is a color;
(2) a color is a musical composition;
(3) therefore, a dog is a musical composition.

Even though the premises do not make sense and the conclusion is false, logicians consider the argument to be valid: if dogs were colors and colors were musical compositions, the conclusion would have to be true. It is for this reason that valid arguments serve to establish conclusions from accepted premises. What makes the argument valid is not the meaning of its words but how they are related: any argument in which the premises and the conclusion are related in the same way is valid. The importance of such structure (referred to as *syntax*) extends to language generally. Knowing the syntax of a language enables you to construct and understand an indefinite number of sentences with the same structure.

In the era before researchers started to invoke ANNs to explain cognitive activities, most researchers assumed that cognition worked much in the manner of logical arguments: an individual was assumed to encode thoughts in structured representations and apply rules that depended on their structure (syntax) to develop new thoughts. This ensured the systematicity of thinking. Since ANNs do not apply rules to structured representations, many theorists, including many philosophers (Fodor & Pylyshyn, 1988), argued they could not account for thinking (see Buckner & Garson, 2019, section 7).

Proponents of ANNs have advanced several responses to this challenge. One is to treat the structures exhibited in thought as patterns to be learned by a neural network realized in the neocortex. Figure 14(b) represents the visual processing system much like a multilayer artificial network. Each brain region shown at higher levels extracts additional patterns from the patterns recognized at lower levels. Brain areas in the central and anterior inferotemporal cortex, at the top of the ventral stream (shown on the right of Figure 14(b) and labeled *inferotemporal stream*), respond to patterns corresponding to abstract categories such as shape, color, or faces. Some areas in this region also respond differentially to categories of objects (e.g., dogs, houses). Barsalou (2008) has suggested how this process of identifying more abstract patterns from more concrete ones can be extended to relational categories such as "on top of" or "a type of."[13] A further challenge is to explain the ability to connect the states in the network that represent these categories in flexible (and sometimes arbitrary) ways to capture systematic relations, as illustrated in the previous example of a valid but nonsensical argument. O'Reilly et al. (2014) have crafted ANNs that can represent many such relations and employ a simulation of the basal ganglia (Section 5.4) to enable flexible combination of these relations with other mental concepts.

To date, models of how the neocortex can implement systematic thought are hypothetical proposals, not grounded in details of neural activity or connectivity. What they show is that it is possible for a structure like the neocortex, assisted by the basal ganglia, to produce systematic cognition. A couple of considerations should be kept in mind, however: human thinking is not perfectly systematic (we make inferential errors) and we often employ other types of reasoning, such as reasoning by analogy and metaphor. Further, many animals that are generally not thought to engage in high-level cognitive reasoning also have an extensive neocortex. The ways in which humans use the neocortex, especially the prefrontal cortex, may reflect in part the cultures in which we live and modes of learning supported by those cultures. One thing these cultures make available are languages, which are themselves powerful representational tools both for logical reasoning and analogical and metaphoric reasoning. The networks in our brains, especially those in the neocortex, have

[13] An important part of Barsalou's project is to show that the categories we use in thinking, including abstract ones, are perceptually grounded.

learned to accommodate the structures available in the languages we acquire, and this may be a significant part of the explanation of our ability to engage in systematic thinking.

9.3 Differences between the Neocortex and Artificial Neural Networks

ANNs provide a powerful framework for modeling important features of the neocortex, but there are significant respects in which the neocortex is different from ANNs. First, in most ANNs, processing connections are only feedforward. Error is propagated backward during learning, but not in processing. Yet, in the neocortex there are at least as many *recurrent projections*, projections from anatomically higher-level processing areas to sensory inputs, as forward projections. Although the full function of these recurrent projections is not understood, they allow a response in a higher-processing region, activated by whatever means, to activate other patterns in the lower-level input areas that are frequently associated with that response.

The recurrent activation of lower layers from higher layers provides an explanation of what is referred to as *top-down* processing, according to which the concepts one applies to stimuli affects how one sees them. These is compelling evidence that we engage in such processing. In a classic experiment, Bruner and Postman (1949) flashed playing cards to participants and asked them to name them. Among the cards were abnormal cards, such as a red four of spades. Participants would regularly report a normal card, for example, a four of hearts, although sometimes noting that something seemed to be wrong with the card (but unable to say what). Feedback from higher visual areas on earlier visual areas overrides the input from the senses. It can also explain abilities such as visually imagining a bird when one hears the word "bird" (Kosslyn, 1994) or reporting seeing features of a bird that were not visible in a given presentation.

The prevalence of recurrent projections in the neocortex has led some neuroscientists (Friston, 2010) and philosophers (Clark, 2013) to advance *predictive coding*, an account of neural processing that reverses the more traditional account that starts from activation of the senses and proceeds to recognition of objects. Instead, these theorists propose that higher processing areas make predictions about subsequent sensory input. For example, if one looks down after viewing a person's face, one expects that one will see a human torso. If the prediction is true, no sensory information is processed and the neural mechanism increases its confidence in making such a prediction in the future. But if the sensory input violates the prediction – one sees the torso of a bear – sensory information is processed further. If violations of expectations are frequent enough, one learns from them and makes different predictions in the future.

A second feature distinguishes processing in the neocortex from that in ANNs: all regions in the neocortex are also interconnected with nuclei in the thalamus and the basal ganglia (Section 5.4), both receiving inputs from them and sending outputs to them. In many cases, these projections form loops that have functional significance. For example, as we have noted in Section 6.3, negative (also positive) feedback loops generate oscillations such as those registered with EEG (Section 3.3). The result is that there are ongoing oscillations in the neocortex at many different frequencies (Buzsáki, 2006). These affect how information is processed (discussed further in Bechtel, 2019). As just one example, when subthreshold oscillations in two different brain regions are synchronized, inputs from one region are more likely to generate action potentials in the other region. Recently, researchers have identified traveling waves – patterns of oscillation that move from region to region in the neocortex – and advanced evidence that these modulate things such as sensitivity to perceptual stimuli (Davis et al., 2020). Another example involves loops that include the basal ganglia, which may be particularly important in controlling processing in the neocortex. As we discussed in Section 5.4, the basal ganglia by default inhibit other brain regions. As a result of loops with the thalamus and neocortex, the basal ganglia decide which activity in the neocortex is released from inhibition and allowed to continue.

9.4 Cognitive Control and the Prefrontal Cortex

An important feature of human cognition is what is referred to as *cognitive control*: the ability to resist habitual or emotionally salient behaviors in order to act in more context-appropriate ways (often ways that are expected to be more beneficial in the long term) (Miller & Cohen, 2001). For example, when a European or North American drives in Thailand, they need to resist their habitual responses to drive on the right side and follow instead Thailand's rule of driving on the left. As another example, if we promised our best friend that we would show up for their party at 11 pm, we might have to suppress our physical and emotional exhaustion to honor our promise. Such activities are common among humans. Even when not required to do so, children often share their candies fairly with their friends, suppressing their desire to eat more themselves.

Cognitive control draws upon the processing capacities of a part of the neocortex that we have not discussed much so far, the prefrontal cortex, which occupies the front part of the frontal lobe (the rear portion primarily contains areas involved in processing motor commands). Like posterior areas of the brain involved in vision, the prefrontal areas comprise multiple different processing areas that have been associated, largely through single-cell recording studies in monkeys and neuroimaging studies in humans (see Section 3.3), with a variety of capacities. These are organized into processing streams that extend those involved in visual processing shown in Figure 14(b). The dorsal "where" or action-oriented stream gives rise to areas that represent actions, often complex actions, and the individual's evaluations of those actions. As one moves forward in the prefrontal cortex, the areas first encountered code for learned associations between sensory stimuli and motor responses. Areas yet further forward code more abstract rules between contexts and classes of actions, including social and moral norms (Carlson & Crockett, 2018) and facilitate thinking about hypothetical actions and future states of the world. In contrast, the continuation of the ventral stream represents increasingly abstract features of objects as well as evaluations of these objects.

Recurrent projections are especially prevalent in the prefrontal cortex. These enable individual areas to maintain active states for prolonged periods after the initiating stimulus has ceased. This provides for a form of temporary memory known as *working memory*, which is taken to be particularly important for carrying out complex actions or actions after brief delays. Goldman-Rakic (1995) demonstrated how animals could retain information in these circuits until needed to perform the action. Fuster, Bodner, and Kroger (2000) emphasize how these circuits support the integration of information from different modalities needed for temporal structuring of behavior. However, the active maintenance of information needs to be coupled with flexible and appropriate update of information to be adaptive – when a social rule is no longer appropriate in guiding behaviors in a particular context, the cognitive system needs to be able to shut down the associated neural activities and replace them with ones that represent the currently appropriate social rules. Interconnections between prefrontal areas and the basal ganglia are important in doing this (Section 5.4).

These capacities, especially the ability of prefrontal areas to encode and maintain active representations of goals, rules, and values, suggest how cognitive control is possible. As these areas are connected to other brain areas, these representations can, for example, activate the relevant sensory inputs, memory representations, and motor outputs needed to perform context-appropriate actions while inhibiting actions that might be triggered directly by sensory stimuli (Miller & Cohen, 2001).

We should note a few features to this account of cognitive control. First, cognitive control involves multiple levels in the various hierarchies in the neocortex: a lower-level motor response can be controlled by contextual information at a higher level, which can in turn be modulated by superordinate contextual information. For example, children are often told to speak in a softer voice while indoors. They often learn quickly, though, that this rule only applies within the presence of an adult. The context of having an adult nearby, then, further contextualizes the indoor rule (Badre & Nee, 2018). Second, in addition to top-down control signals traveling from anatomically higher levels to lower levels in a given information-processing hierarchy, recent empirical literature reveals control signals that are bottom-up and lateral, enabling controllers across different levels and hierarchies to constrain each other (Cisek & Kalaska, 2010). We turn to violations of hierarchical control organization in the next section.

9.5 Summary

The cortex, especially the neocortex, is organized differently from the rest of the brain. Within the neocortex, there is a hierarchy of processing areas in which neurons in one area project to those in subsequent ones. Artificial neural networks, modeled on this pattern of organization, suggest that the neocortex is a powerful pattern recognition system. We sketched how this structure can account for the systematicity exhibited in human cognition. We also emphasized the importance of recurrent projections in the neocortex and the connections of regions throughout the neocortex with subcortical structures, especially the thalamus and basal ganglia, and sketched how these enable humans to exert cognitive control.

10 How Is the Whole Nervous System Organized?

In Section 6.5, we introduced the notion of control mechanisms and noted that the nervous system consists of control mechanisms. In Section 7.3, we raised the issue of how control mechanisms, in general, are organized. In this section, we focus on how neural control mechanisms, in particular, are organized, considering two alternatives: organization in hierarchical pyramids or into heterarchical networks.

10.1 Hierarchical Pyramid Organization

In many social systems, such as corporations and the military, control is organized hierarchically in a pyramid, as in Figure 24(a). In this arrangement, multiple local controllers report to a smaller number of controllers at a higher level. This is iterated, culminating in a chief executive, a president, or a general, with whom the buck stops (to use an expression coined by US President Harry

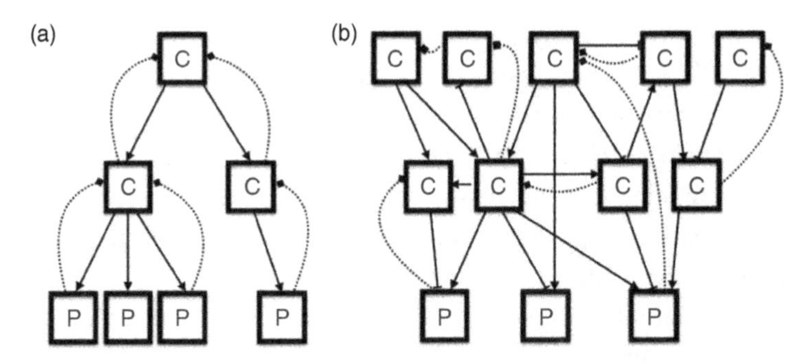

Figure 24 (a) A typical hierarchical pyramid of control processes with information (dotted arrows) directed toward higher layers and commands directed to lower levels. Sensory inputs are represented by dotted arrows. Arrows and edge-ended lines indicate excitatory and inhibitory control. (b) Heterarchical organization that violates several features of a hierarchical pyramid.

Truman). In the control hierarchy, the lower-level control mechanisms are, adopting Dennett's (1991) term, bureaucratic. They function to provide information for the central control mechanism or to work out the implementation of its commands. On such a scheme, there are in principle no conflicts between lower-level and central controllers, since the lower-level controllers do not have their own agenda and function solely to serve the central controller.

The nervous system is often conceptualized as organized as a hierarchical pyramid. Local nerve nets and pattern generators are brought under the control of individual ganglia (Section 2.3). As a central brain evolved, peripheral ganglia were brought under the control of more central ones such as the nuclei in the vertebrate brain. Then, as the neocortex evolved, it assumed control: sensory inputs are fed up to it and it sends motor commands back down to subordinate levels of control. Except for the positioning of the basal ganglia and thalamus, Figure 15 presents such a picture.

One consequence of the pyramid structure is that entities at each level in the hierarchy face greater demands. In order to direct effectively a company or a military structured this way, the chief executive needs to acquire all of the relevant information and use it to make decisions. Pyramid organization is also conservative, as innovation is permitted only at the top. As a result, some social institutions relax the hierarchical pyramid, delegating to lower-level controllers the ability to act independently. Coordination between controllers then becomes a challenge. To some degree, this can be achieved by direct interaction of controllers at a given level (as between ganglia in the leech example discussed in Section 2.3). A further departure from the pyramid organization is to forego the strict layering of levels, allowing information and commands to skip layers or to be asserted within layers. At the extreme, there is no executive at the top of the pyramid and there are more control agents as one moves up levels (Figure 24(b)). McCulloch (1945) introduced the term *heterarchy* for situations in which one's preferences are not organized hierarchically (e.g., one prefers A to B, B to C, and C to A), and Pattee (1991) extended it to control systems in living organisms that significantly depart from the hierarchical pyramid organization.

10.2 Heterarchical Network Organization

The neocortex is generally understood to be organized hierarchically. But given its multiple processing streams (illustrated in Figure 14(b)), it does not conform to the pyramid structure. When we turn to the nervous system more generally, departures from the hierarchical pyramid mode of organization become more notable. This is evident when we consider some of the organisms we discussed in other sections. In the jellyfish (Section 2.2), multiple different control mechanisms, each responsive to conditions that require alterations in the default mode of swimming, act on the same nerve network. The various ganglia in the leech (Section 2.3) and in *C. elegans* (Section 4.2) act in relative independence. Despite becoming colocated in the brain over the course of evolution, the ANS and BNS (Section 2.3) continue to operate independently to a significant extent. Regions of the hypothalamus, the arcuate nucleus, the lateral zone, and the SCN each regulate different behaviors (feeding, attention versus sleep, circadian rhythms, respectively). The neocortex cannot dictate circadian rhythms (except by directing actions in the world such as exposing oneself to daylight).

How do control structures in the brain become organized in a heterarchical network? For a clue, consider one of the principles we discussed in Section 6.4: in many real world networks, degree (number of connections per node) is distributed accordingly to a power law (Section 6.4) in which a few nodes are extremely highly connected. Barabási and Bonabeau (2003) proposed a process by which networks come to exhibit this feature: in many contexts, when an edge is added from a given node to another, it is more likely that it will connect to an already highly connected node. We can couple this with a further factor – in control networks, new edges are introduced to achieve better control, especially to overcome a shortcoming of the current control system, and this often involves connecting to another control mechanism. There is a parallel process in revising computer software. Unlike in organisms, the initial code of a computer program may have been intelligently designed. But no matter how intelligently designed and well tested it is, it will likely fail in some contexts. When it does so, programmers do not redesign from scratch, but patch the current code by adding whatever new code will address the situation without obviously compromising other parts of the program (these are often referred to as *kludges*). Biological networks, including neural networks, evolve in the same manner – retaining new connections that happen to form between existing components when they will improve the organism's performance in the current environment. The check on these new connections is much like the check on software kludges – does the additional connection enhance the ability of the organism to maintain itself (or at least do not render it much less likely to survive)?

In Figure 24, we presented the heterarchical alternative to the hierarchical pyramid in terms of layers arranged hierarchically. A reason for maintaining a representation in terms of layers is that individual control mechanisms operate on other specific mechanisms and this seems well represented by putting a control mechanism at a higher level than the mechanism it controls. But the hierarchy is already breaking down with the inclusion of edges between nodes in the same layer and between nodes two layers apart. Adopting the perspective that evolution adds kludges to an existing control system, such departures from hierarchy are to be expected as there is no principled reason to maintain hierarchical organization. One could also add connections so that lower-level nodes act to control those at a higher level. Ultimately, control systems, especially those that have evolved, are better represented as networks.

10.3 How to Achieve Coherent and Intelligent Control in a Heterarchical Network

One reason many people find the hierarchical pyramid structure to be intuitive is that it solves two important issues of control: coherence and intelligence. The issue of coherence concerns how neural control mechanisms manage to produce more or less coherent behaviors. Having a central executive making the call with other lower-level controllers merely implementing the details certainly seems like a good way to ensure coherence of behaviors. The issue of intelligence concerns how control mechanisms can generate adaptive behaviors in a wide range of novel contexts. As we discussed earlier, control mechanisms need to adjust their basic mechanisms to operate appropriately in a situation. Then, these different basic mechanisms can work together to produce adaptive behaviors. But how do control mechanisms know the appropriate control signals, especially for novel situations they have not encountered before? The central executive is supposedly the source of intelligence. It is often assumed to possess rich information about the world and to operate on this information in order to exercise context-appropriate control over more basic mechanisms.

However, the brain does not appear to have a hierarchical pyramid structure. This requires us to address the question: How can heterarchical networks achieve coherent and intelligent control of organisms?

Theorists have identified strategies that enhance coherence between multiple control mechanisms (Clark, 2014). For example, the fact that multiple controllers all confront and get feedback from the same external world promotes coherence. Also, various neuromodulators and hormones dissipate broadly in neural and bodily systems, communicating information about conditions inside and outside the organism with different control and production mechanisms. Further, local communications between controllers can reduce conflicts between the decisions made by different controllers. We noted this with the leech (Section 2.3): even though the decision to walk or swim is made in each ganglion, the control mechanisms in different ganglia act on each other so that they arrive at a coherent action. Finally, of special importance in promoting coherence are the basal ganglia, as they provide a common structure in which evaluations made by multiple controllers distributed across the brain are brought together and outputs are sent back to these different controllers, modulating their individual behaviors. By forcing the integration of these initially diverse inputs, the basal ganglia have the infrastructure to enhance the coherence of the control decisions made by different brain systems.

Outside of neuroscience, there are numerous examples of how intelligent behaviors can emerge from a network of controllers, each of which has only partial access to the information about the world. Marvin Minsky (1986), a pioneer in artificial intelligence, offered the metaphor of a "society of mind" that fits well with the view of the nervous system as a network of heterarchically organized control mechanisms.[14] Could a society give rise to a self that acts in a unified, intelligent way? Examples such as honeybee swarms suggest that integrated intelligent behavior can arise from collective activity.[15] Modern democratic societies embrace the idea of determining courses of action through voting. In the following, we sketch how a heterarchical nervous system could produce intelligent behaviors by operating like a political democracy.[16]

Appealing to what is called the "wisdom of the crowd effect," social choice theory argues that under the right conditions, aggregating the decisions of multiple individuals (e.g., through voting) is more likely to result in a correct

---

[14] We develop the society of mind metaphor in terms of control – the human mind is composed of a massive number of control mechanisms. For a development in terms of representations, see Rupert (2011).

[15] For a relevant discussion on whether we could treat different types of distributed systems, such as bee swarms or human society, as cognitive systems, see Huebner (2014).

[16] For a more detailed account, see L. T. Huang (2017). Neurodemocracy: Self-organization of the embodied mind (Ph.D. dissertation), University of Sydney, http://hdl.handle.net/2123/16845.

decision than relying on an individual's decisions. This can be demonstrated mathematically (List, 2013). Consider three people, each of whom has an independent reliability of 0.8 of getting the answer right. If the three people vote and go with the majority, the reliability increases to 0.896.[17] The likelihood of correctness increases further as more people are included. This result can be generalized to a wide range of conditions. An explanation for this is that the collective is integrating relatively reliable information stemming from different sources. This result applies as well to the nervous system. Multiple nuclei can integrate different sources of information, arriving collectively at a more reliable decision than if the organism relied on just one. In the vertebrate nervous system, the basal ganglia are organized to determine the winner in a manner comparable to voting: the inputs represent assessments of different alternatives and the competition to control the direct and indirect pathways culminates in a Go/NoGo decision. Thus, there is reason to think that a heterarchical system that makes decisions through processes such as voting could give rise to coherent and intelligent agency. If it can, there is less reason to assume that control mechanisms must be organized hierarchically.

10.4 Summary

Accepting that neural systems operate as control mechanisms, in this section we have considered two alternative patterns of organization: a hierarchical pyramid or a heterarchical network. Despite the challenge of maintaining coherence and achieving intelligence in a heterarchical scheme, the nervous system does appear to be organized heterarchically. We have briefly considered how coherent and intelligent actions might be generated in a heterarchically organized system.

11 What Does Neuroscience Teach Us about Who We Are?

We conclude this Element by considering a neurophilosophical question (Section 1): What does the knowledge acquired in neuroscience, some of which we have reviewed in this Element, tell us about ourselves? When most people are asked to characterize themselves, they begin with traits such as race, sex, gender, age, height, hair color, and so on. As biological organisms, they appeal in part to their history – they were born at a certain moment from specific parents and have followed a trajectory across the earth. Many will refer to important events in their personal history, their job or profession, and preferred activities. But beyond that, many people think that these are just the external

[17] The result is reached by adding up the probabilities of four different scenarios where the group gets the answer right ($0.8 \times 0.8 \times 0.8 + 0.8 \times 0.8 \times 0.2 + 0.8 \times 0.2 \times 0.8 + 0.2 \times 0.8 \times 0.8 = 0.896$).

expressions of something internal: their self. On the Temple of Apollo at Delphi in Greece was inscribed the injunction: "Know thyself." Given the role of the nervous system in controlling behavior, part of following the Oracle's injunction might be to consider who we are in light of what is known about our nervous system.

If one views the nervous system as a hierarchical control system with a central executive issuing commands, the executive implemented in the prefrontal cortex might be the self we need to know. As we have seen throughout this Element, however, the nervous system is organized heterarchically, with relatively independent control mechanisms located in many different parts of the nervous system. Activities such as eating, sleeping, and reproducing are controlled by nuclei in the hypothalamus. These are important features of who we are. Memories, especially memories for events in our lives, are developed in the hippocampus and ultimately laid down in the neocortex. Who we are is often revealed in our choice of actions, in which the basal ganglia play a major role. If the nervous system is really organized heterarchically, what sense can be made of *a* self to be found in our nervous system?

11.1 Reporting on Our Mental Lives

In this section, we explore what might seem a radical hypothesis: that there is not a self to learn about; rather each of us, drawing upon our ability to use language, constructs one. The key idea draws from Wilfred Sellars' (1956) Myth of Jones, which he offered, not as an historical account, but as a means to illustrate the status of our reports on our own mental states. According to his myth, before people developed the ability to describe their own mental states, they had developed natural science and made successful predictions about entities in the world. To explain how organisms behaved, they posited mechanisms within them. One scientist, Jones, turned this technique on human behavior: he hypothesized inner entities he called *thoughts* and developed a theoretical framework with which he could successfully predict how different people would behave on the basis of the thoughts he attributed to them. Other people learned to use this theoretical framework and even applied it to themselves, initially inferring what they thought based on their overt behaviors. Then Jones taught Dick to describe his thoughts without first consulting his overt behavior in much the way neuroscientists train nonhuman subjects to carry out tasks in their experiments – by giving him positive when his self-ascriptions fit those Jones made based on Dick's behavior and negative feedback when they didn't. Dick successfully learned to do this; although he had no idea how he does so, he reports on his own decision making in terms of his thoughts. Sellars' point is that although there must be some basis on which Dick succeeds in doing this, there is no need to view him as reporting or "introspecting" internal states. Rather, he has learned to extract patterns, which, as we saw, is the forte of the neocortex.[18]

The theoretical framework Jones developed and Dick learned corresponds to what philosophers refer to as *folk psychology*. It characterizes humans in terms of attitudes toward what are often referred to as *propositions*. The idea is that in one's mental life, one represents information in propositions that, for our purposes, we can treat as statements in a language such as English. An example proposition would be "My nextdoor neighbor has a cat." According to folk psychology, one can adopt different attitudes toward this proposition – one can believe that it is true, doubt that it is true, fear that it might be true, wish that it were true, and so on. Further, one can reason in terms of these propositional attitudes: from the propositional attitudes of believing "there is yogurt in my refrigerator," "I would like to eat yogurt," and "if I go to my refrigerator, I can eat what is in it," one can infer "I should go to my refrigerator." On Sellars' account, one need not treat folk psychological statements as describing neural events. They are constructs in a story we tell about other people and ourselves. Nonetheless, in terms of them, we can provide useful accounts of how we and others behave. Moreover, we can update these accounts when they go astray and make better predictions in the future.[19] For example, if you think your friend believes it is going to rain, wants to stay dry, and believes bringing her umbrella will enable her to stay dry, you can infer that she will have her umbrella with her. If she shows up without it, you can inquire whether she did not believe it was going to rain, did not want to stay dry, or was not acting rationally.

11.2 Making Norms for Action Explicit and Living by Them

As we discussed in Section 6.5, control mechanisms can be viewed as implementing norms through their response to the measurements that they make. In the case of the Watt governor, the norm that is implemented is built into its design – it was designed to maintain a steam engine at a specific speed. The various control mechanisms in the brain likewise implement norms that are incorporated in them either through evolution or through learning. When we think of norms, we often focus on moral norms. These are norms that we take to be capable of being expressed in language, rationally discussed, and chosen, at which point they influence one's actions. How do these moral norms relate to those implemented in neural control mechanisms in organisms?

[18] This analysis is developed further in Bechtel (2008, chapter 7). For recent philosophical and scientific development along this line, see Schwitzgebel (2019).

[19] Eliminativists such as Paul M. Churchland (1981) see the failures of folk psychology to make correct predictions as a reason to repudiate it as a false theory.

The account of how we can report on our mental lives sketched in the previous section provides a framework for representing our mental life in language, but it does not directly address how the results of explicit adoption of norms can become efficacious. One might infer that if accounts of mental processes were constructions, they could not be efficacious. But this is wrong: we construct our account of our mental activities using our brains. As we construct such an account, we alter processes in our brains. This applies as well to our discourse about norms. As we adopt norms and remind ourselves of them, they may direct our behavior (Frankish, 2004). At present, we know very little about the brain processes that figure in these activities and so are not in a position to spell out how they affect behavior. But given their reliance on language, it seems reasonable to assume that linguistic processes (e.g., generating inner speech) can affect brain mechanisms that are engaged in the selection of behavior.

We are aware that sometimes even when we commit ourselves explicitly to a given norm, we will violate it. We commit ourselves to leaving the party by 11 pm, but end up staying to 1 am. Philosophers characterize this as *weakness of the will*. Given what we have said about the heterarchical organization of the brain, this phenomenon is not surprising. Other neural control mechanisms compete and win out. But that does not mean explicit commitments to norms are always inefficacious. As emphasized by many moral theorists, one way to make our moral commitments efficacious is to turn them into habits by, for example, giving ourselves a reward when we fullfil our commitments repeatedly. Another is to draw attention to our commitments. If you commit yourself to a norm publicly, that might strengthen that norm when it is in competition with others. Or others may remind us of our commitment (for a recent development of the mind-shaping effect of the social practice of articulating our values, see McGeer, 2015).

11.3 Constructing a Self

What should we make of the Delphic Oracle's injunction in light of our discussions of heterarchy and the constructive character of our accounts of our mental lives? On the account we have offered, one's concept of oneself would also be a construct.

Out of what do we construct ourselves? For most people, memories of episodes in our past are important elements. Tulving (1983) not only coined the term *episodic memory* for these memories but characterized them as mental time travel, thereby capturing how recalling an event seems like reliving it. Although at least with vivid memories people often have the sense of simply rehearsing the past, there is compelling evidence that recalling a memory does not involve retrieving a record of the past but rather reconstructing the past event from multiple sources of information. Such reconstruction is often affected by what happened on previous occasions during which one recalled the event (as shown by Loftus, 1975, a detail suggested by someone else during the recall can be subsequently remembered as part of the initial event). This ability to put material together into a past narrative also enables us to project ourselves into the future, characterizing ourselves in terms of what we hope to become. Beyond memories, we also characterize ourselves in terms of traits and abilities that we take ourselves to have and norms we hope to uphold.

If our self is a construct, then perhaps the meaning of the Delphic Oracle is that each of us needs to construct our self-concept, one that presents a narrative of our past, projects ourself into the future, and frames who we are. We can draw upon it in making major life decisions. Those decisions will also contribute to constructing our self-concept in the future. Unfortunately, our self-concept will not always be effective in guiding the decisions we make. Our nervous system is still heterarchical, and control mechanisms, whether linked to our self-concept or not, will generate many of our actions more or less independently. But, like the moral norms we discussed in the previous section, our self-concept can be invoked to direct and constrain the decisions we make, thereby providing focus to our lives.

References

Abrahamsen, A., Sheredos, B., & Bechtel, W. (2018). Explaining visually using mechanism diagrams. In S. Glennan & P. Illari (Eds.), *The Routledge handbook of mechanisms* (pp. 238–254). London: Routledge.

Akins, K. (1996). On sensory systems and the "aboutness" of mental states. *The Journal of Philosophy, 93*(7), 337–372.

Andersen, R. A., Essick, G. K., & Siegel, R. M. (1985). Encoding of spatial location by posterior parietal neurons. *Science, 230*, 456–458.

Anderson, M. L. (2014). *After phrenology: Neural reuse and the interactive brain*. Cambridge, MA: MIT Press.

Ankeny, R. A., & Leonelli, S. (2020). *Model organisms*. Cambridge: Cambridge University Press.

Ardiel, E. L., & Rankin, C. H. (2010). An elegant mind: Learning and memory in *Caenorhabditis elegans. Learning & Memory, 17*(4), 191–201. DOI:10.1101/lm.960510.

Arendt, D., Tosches, M. A., & Marlow, H. (2016). From nerve net to nerve ring, nerve cord and brain–evolution of the nervous system. *Nature Reviews Neuroscience, 17*(1), 61–72. DOI:10.1038/nrn.2015.15.

Arkett, S. A., Mackie, G. O., & Meech, R. W. (1988). Hair cell mechanoreception in the jellyfish *Aglantha digitale*. *Journal of Experimental Biology, 135*, 329–342.

Badre, D., & Nee, D. E. (2018). Frontal cortex and the hierarchical control of behavior. *Trends in Cognitive Sciences, 22*(2), 170–188. DOI:10.1016/j.tics.2017.11.005.

Barabási, A.-L., & Bonabeau, E. (2003). Scale-free networks. *Scientific American, 288*(5), 60–69.

Bargmann, C. I., & Marder, E. (2013). From the connectome to brain function. *Nature Methods, 10*(6), 483–490. DOI: 10.1038/Nmeth.2451.

Barsalou, L. W. (2008). Cognitive and neural contributions to understanding the conceptual system. *Current Directions in Psychological Science, 17*(2), 91–95. DOI: 10.1111/j.1467–8721.2008.00555.x.

Bechtel, W. (2008). *Mental mechanisms: Philosophical perspectives on cognitive neuroscience*. London: Routledge.

Bechtel, W. (2009). Molecules, systems, and behavior: Another view of memory consolidation. In J. Bickle (Ed.), *Oxford handbook of philosophy and neuroscience* (pp. 13–40). Oxford: Oxford University Press.

Bechtel, W. (2016). Investigating neural representations: The tale of place cells. *Synthese*, *193*(5), 1287–1321. DOI:10.1007/s11229-014-0480-8.

Bechtel, W. (2021). Explaining features of fine-grained phenomena using abstract analyses of phenomena and mechanisms: Two examples from chronobiology. *Synthese*, *198*(1), 1–23. DOI:10.1007/s11229-017-1469-x.

Bechtel, W. (2019). Networks and dynamics: 21st century neuroscience. In S. Robins, J. Symons, & P. Calvo (Eds.), *The Routledge companion to philosophy of psychology* (pp. 456–470). London: Routledge.

Bechtel, W., & Abrahamsen, A. (2002). *Connectionism and the mind: Parallel processing, dynamics, and evolution in networks* (second ed.). Oxford: Blackwell.

Bechtel, W., & Abrahamsen, A. (2005). Explanation: A mechanist alternative. *Studies in History and Philosophy of Biological and Biomedical Sciences*, *36* (2), 421–441.

Bechtel, W., & Abrahamsen, A. (2010). Dynamic mechanistic explanation: Computational modeling of circadian rhythms as an exemplar for cognitive science. *Studies in History and Philosophy of Science Part A*, *41*(3), 321–333.

Bechtel, W., & Richardson, R. C. (1993/2010). *Discovering complexity: Decomposition and localization as strategies in scientific research*. Princeton, NJ: /Princeton, NJ:Princeton University Press (1993)/ Cambridge, MA: MIT Press (2010).

Berger, H. (1930). Über das Elektrenkephalogramm des Menschen. Zweite Mitteilung. *Journal für Psychologie und Neurologie*, *40*, 160–179.

Bickle, J. (2006). Reducing mind to molecular pathways: Explicating the reductionism implicit in current cellular and molecular neuroscience. *Synthese*, *151*, 411–434.

Bjursten, L. M., Norrsell, K., & Norrsell, U. (1976). Behavioural repertory of cats without cerebral cortex from infancy. *Experimental Brain Research*, *25* (2), 115–130. DOI:10.1007/BF00234897.

Bogacz, R., & Gurney, K. (2007). The basal ganglia and cortex implement optimal decision making between alternative actions. *Neural Computation*, *19*(2), 442–477. DOI:10.1162/neco.2007.19.2.442.

Briggman, K. L., Abarbanel, H. D. I., & Kristan, W. B. (2005). Optical imaging of neuronal populations during decision-making. *Science*, *307*(5711), 896–901. DOI:10.1126/Science.1103736.

Britten, K. H., Shadlen, M. N., Newsome, W. T., & Movshon, J. A. (1992). The analysis of visual motion: A comparison of neuronal and psychophysical performance. *The Journal of Neuroscience*, *12*, 4745–4765.

Broca, P. (1861). Remarques sur le siége de la faculté du langage articulé, suivies d'une observation d'aphemie (perte de la parole). *Bulletin de la Société Anatomique*, *6*, 343–357.

Brodmann, K. (1909/1994). *Localization in the cerebral cortex* (L. J. Garvey, trans.). New York: Springer.

Brook, A. (2009). Introduction: Philosophy in and philosophy of cognitive science. *Topics in Cognitive Science*, *1*(2), 216–230. DOI:10.1111/j.1756-8765.2009.01014.x.

Bruner, J. S., & Postman, L. (1949). On the perception of incongruity: A paradigm. *The Journal of Personality*, *18*, 206–223.

Buchwald, J. S., & Brown, K. A. (1973). Subcortical mechanisms of behavioral plasticity. In J. D. Maser (Ed.), *Efferent organization and the integration of behavior* (pp. 99–136). New York: Academic Press.

Buckner, C., & Garson, J. (2019). Connectionism. In E. N. Zalta (Ed.), *The Stanford encyclopedia of philosophy*. Retrieved from https://plato .stanford.edu/archives/fall2019/entries/connectionism.

Burk, J. A., & Fadel, J. R. (Eds.). (2019). *The orexin/hypocretin system: Functional roles and therapeutic potential*. New York: Academic Press.

Burnston, D. C. (2016). A contextualist approach to functional localization in the brain. *Biology & Philosophy*, *31*(4), 527–550. DOI:10.1007/s10539-016-9526-2.

Buzsáki, G. (2006). *Rhythms of the brain*. Oxford: Oxford University Press.

Buzsáki, G. (2010). Neural syntax: cell assemblies, synapsembles, and readers. *Neuron, 68*(3), 362–385. doi:10.1016/j.neuron.2010.09.023

Carlson, R. W., & Crockett, M. J. (2018). The lateral prefrontal cortex and moral goal pursuit. *Current Opinion in Psychology*, *24*, 77–82. DOI:10.1016/ j.copsyc.2018.09.007.

Chakravarthy, V. S., & Balasubramani, P. P. (2018). The basal ganglia system as an engine for exploration. In V. S. Chakravarthy & A. A. Moustafa (Eds.), *Computational neuroscience models of the basal ganglia* (pp. 59–96). New York: Springer. DOI:10.1007/978-981-10-8494-2_5.

Chalfie, M., Sulston, J. E., White, J. G., Southgate, E., Thomson, J. N., & Brenner, S. (1985). The neural circuit for touch sensitivity in *Caenorhabditis elegans*. *Journal of Neuroscience*, *5*(4), 956–964.

Chemero, A. (2000). Anti-representationalism and the dynamical stance. *Philosophy of Science*, *67*(4), 625–647. DOI:10.1086/392858.

Chemero, A., & Silberstein, M. (2008). After the philosophy of mind: Replacing scholasticism with science. *Philosophy of Science*, *75*(1), 1–27. DOI:10.1086/587820.

Churchland, P. M. (1981). Eliminative materialism and propositional attitudes. *The Journal of Philosophy*, *78*, 67–90.

Churchland, P. S. (1986). *Neurophilosophy: Toward a unified science of the mind-brain*. Cambridge, MA: MIT Press/Bradford Books.

Cisek, P. (2019). Resynthesizing behavior through phylogenetic refinement. *Attention, Perception, & Psychophysics*, doi:10.3758/s13414-019-01760-1.

Cisek, P., & Kalaska, J. F. (2010). Neural mechanisms for interacting with a world full of action choices. *Annual Review of Neuroscience, 33*, 269–298. DOI:10.1146/annurev.neuro.051508.135409.

Clark, A. (2013). Whatever next? Predictive brains, situated agents, and the future of cognitive science. *The Behavioral and Brain Sciences, 36*(3), 181–204. DOI:10.1017/S0140525X12000477.

Clark, A. (2014). *Mindware: An introduction to the philosophy of cognitive science* (second ed.). New York: Oxford University Press.

Colaço, D. (2018). Rip it up and start again: The rejection of a characterization of a phenomenon. *Studies in History and Philosophy of Science Part A, 72*, 32–40. DOI:10.1016/j.shpsa.2018.04.003.

Colgin, L. L., Moser, E. I., & Moser, M.-B. (2008). Understanding memory through hippocampal remapping. *Trends in Neurosciences, 31*(9), 469–477. DOI:10.1016/j.tins.2008.06.008.

Corkin, S. (2013). *Permanent present tense: The unforgettable life of the amnesic patient, H.M.* New York: Basic Books.

Craver, C. F. (2003). The making of a memory mechanism. *Journal of the History of Biology, 36*, 153–195.

Craver, C. F. (2006). When mechanistic models explain. *Synthese, 153*(3), 355–376. DOI:10.1007/s11229-006-9097-x.

Craver, C. F. (2007). *Explaining the brain: Mechanisms and the mosaic unity of neuroscience*. New York: Oxford University Press.

Craver, C. F. (2008). Physical law and mechanistic explanation in the Hodgkin and Huxley model of the action potential. *Philosophy of Science, 75*(5), 1022–1033. DOI:10.1086/594543.

Craver, C. F., & Darden, L. (2013). *In search of mechanisms: Discoveries across the life sciences*. Chicago: University of Chicago Press.

Craver, C. F., & Tabery, J. (2019). Mechanisms in science. In E. N. Zalta (Ed.), *The Stanford excyclopedia of philosophy*. Retrieved from https://plato.stanford.edu/archives/sum2019/entries/science-mechanisms/.

Crisp, K. M., & Mesce, K. A. (2006). Beyond the central pattern generator: Amine modulation of decision-making neural pathways descending from the brain of the medicinal leech. *Journal of Experimental Biology, 209*(9), 1746–1756. DOI:10.1242/jeb.02204.

Cushing, H. (1909). A note upon the faradic stimulation of the post-central gyrus in conscious patients. *Brain*, *32*, 44–53.

Damasio, A. R. (1995). *Descartes' error*. New York: G. P. Putnam.

Davis, Z. W., Muller, L., Martinez-Trujillo, J., Sejnowski, T., & Reynolds, J. H. (2020). Spontaneous travelling cortical waves gate perception in behaving primates. *Nature*, *587*(7834), 432–436. DOI:10.1038/s41586-020-2802-y.

Dennett, D. C. (1991). *Consciousness explained*. New York: Little, Brown.

Diba, K., & Buzsáki, G. (2007). Forward and reverse hippocampal place-cell sequences during ripples. *Nature Neuroscience*, *10*(10), 1241–1242. DOI:10.1038/nn1961.

Egan, F. (2019). The nature and function of content in computational models. In M. Sprevak & M. Colombo (Eds.), *The Routledge handbook of the computational mind* (pp. 247–258). London: Routledge.

Eichenbaum, H. (2002). *The cognitive neuroscience of memory: An introduction*. Oxford: Oxford University Press.

Fodor, J. A., & Pylyshyn, Z. W. (1988). Connectionism and cognitive architecture: A critical analysis. *Cognition*, *28*, 3–71.

Fowler, O. S. (1890). *The illustrated self-instructor in phrenology and physiology*. New York: Fowler and Wells.

Frankish, K. (2004). *Mind and supermind*. Cambridge: Cambridge University Press.

Friston, K. (2010). The free-energy principle: A unified brain theory? *Nature Reviews Neuroscience*, *11*(2), 127–138. DOI:10.1038/nrn2787.

Fuster, J. M., Bodner, M., & Kroger, J. K. (2000). Cross-modal and cross-temporal association in neurons of frontal cortex. *Nature*, *405*(6784), 347–351. DOI:10.1038/35012613.

Gall, F. J. (1812). *Anatomie et physiologie du systême nerveaux et général, et du cerveau en particulier, avec des observations sur la possibilité de reconnoitre plusieurs dispositions intellectuelles et morales de l'homme et des animaux, par la configuration de leur têtes*. Paris: F. Schoell.

Galvani, L. (1791). *De viribus electricitatis in motu musculari commentarius*. Bologna: Ex typographia Instituti Scientiarum.

Gaudry, Q., Ruiz, N., Huang, T., Kristan, W. B., III, & Kristan, W. B., Jr. (2010). Behavioral choice across leech species: chacun à son goût. *The Journal of Experimental Biology*, *213*(8), 1356–1365. DOI:10.1242/jeb.039495.

Gibson, J. J. (1979). *The ecological approach to visual perception*. Boston: Houghton Mifflin.

Goldbeter, A. (1995). A model for circadian oscillations in the *Drosophila* period protein (PER). *Philosophical Transactions of the Royal Society B: Biological Sciences*, *261*(1362), 319–324. DOI:10.1098/rspb.1995.0153.

Goldman-Rakic, P. S. (1995). Cellular basis of working memory. *Neuron*, *14*(3), 477–485. DOI:10.1016/0896-6273(95)90304-6.

Grillner, S., & El Manira, A. (2019). Current principles of motor control, with special reference to vertebrate locomotion. *Physiological Reviews*, *100*(1), 271–320. DOI:10.1152/physrev.00015.2019.

Gross, C. G., Rocha-Miranda, C. E., & Bender, D. B. (1972). Visual properties of neurons in inferotemporal cortex of the macaque. *Journal of Neurophysiology*, *35*, 96–111.

Haken, H., Kelso, J. A. S., & Bunz, H. (1985). A theoretical model of phase transitions in human hand movements. *Biological Cybernetics*, *51*(5), 347–356.

Hardin, P. E., Hall, J. C., & Rosbash, M. (1990). Feedback of the *Drosophila period* gene product on circadian cycling of its messenger RNA levels. *Nature*, *343*(6258), 536–540.

Haselager, W. F. G., de Groot, A., & van Rappard, H. (2003). Representationalism vs. anti-representationalism: A debate for the sake of appearance. *Philosophical Psychology*, *16*, 5–23.

Hazy, T. E., Frank, M. J., & O'Reilly, R. C. (2007). Towards an executive without a homunculus: Computational models of the prefrontal cortex/basal ganglia system. *Philosophical Transactions of the Royal Society B: Biological Sciences*, *362*(1485), 1601–1613. DOI:10.1098/rstb.2007.2055.

Hempel, C. G. (1965). *Aspects of scientific explanation and other essays in the philosophy of science*. New York: The Free Press.

Hendricks, J. C., Finn, S. M., Panckeri, K. A., et al. (2000). Rest in *Drosophila* is a sleep-like state. *Neuron*, *25*(1), 129–138.

Hills, T. T. (2006). Animal foraging and the evolution of goal-directed cognition. *Cognitive Science*, *30*(1), 3–41. DOI:10.1207/s15516709cog0000_50.

Hodgkin, A. L., & Huxley, A. F. (1952). A quantitative description of membrane current and its application to the conduction and excitation of nerve. *Journal of Physiology*, *117*, 500–544.

Hubel, D. H., & Wiesel, T. N. (1959). Receptive fields of single neurones in the cat's striate cortex. *Journal of Physiology*, *148*, 574–591.

Huebner, B. (2014). *Macrocognition: A theory of distributed minds and collective intentionality*. Oxford: Oxford University Press.

Huneman, P. (2010). Topological explanations and robustness in biological sciences. *Synthese*, *177*(2), 213–245. DOI:10.1007/s11229-010-9842-z.

Huneman, P. (2018). Diversifying the picture of explanations in biological sciences: Ways of combining topology with mechanisms. *Synthese*, *195*(1), 115–146. DOI:10.1007/s11229-015-0808-z.

Joiner, W. J. (2016). Unraveling the evolutionary determinants of skoizleep. *Current Biology, 26*(20), R1073–R1087. DOI:10.1016/j.cub.2016.08.068.

Kanwisher, N., McDermott, J., & Chun, M. M. (1997). The fusiform face area: A module in human extrastriate cortex specialized for face perception. *Journal of Neuroscience, 17*(11), 4302–4311.

Keene, A. C., & Duboue, E. R. (2018). The origins and evolution of sleep. *The Journal of Experimental Biology, 221*(11), jeb159533. DOI:10.1242/jeb.159533.

Keijzer, F., van Duijn, M., & Lyon, P. (2013). What nervous systems do: Early evolution, input–output, and the skin brain thesis. *Adaptive Behavior, 21*(2), 67–85. DOI:10.1177/1059712312465330.

Koizumi, O. (2016). Origin and evolution of the nervous system considered from the diffuse nervous system of Cnidarians. In S. Goffredo & Z. Dubinsky (Eds.), *The Cnidaria, past, present and future: The world of Medusa and her sisters* (pp. 73–91). Cham: Springer International Publishing.

Konopka, R. J., & Benzer, S. (1971). Clock mutants of *Drosophila melanogaster. Proceedings of the National Academy of Sciences of the United States of America, 89*, 2112–2116.

Kosslyn, S. M. (1994). *Image and brain: The resolution of the imagery debate.* Cambridge, MA: MIT Press.

Leng, G. (2018). *The heart of the brain: The hypothalamus and its hormones.* Cambridge, MA: MIT Press.

Levy, A. (2013). What was Hodgkin and Huxley's achievement? *The British Journal for the Philosophy of Science, 65*(3), 469–492. DOI:10.1093/bjps/axs043.

Levy, A., & Bechtel, W. (2013). Abstraction and the organization of mechanisms. *Philosophy of Science, 80*(2), 241–261. DOI:10.1086/670300.

List, C. (2013). Social choice theory. In E. N. Zalta (Ed.), *The Stanford encyclopedia of philosophy.* Retrieved from https://plato.stanford.edu/archives/win2013/entries/social-choice.

Loftus, E. F. (1975). Leading questions and the eyewitness report. *Cognitive Psychology, 7*, 550–572.

Machamer, P., Darden, L., & Craver, C. F. (2000). Thinking about mechanisms. *Philosophy of Science, 67*, 1–25.

Mackie, G. O. (2004). Central neural circuitry in the jellyfish *Aglantha* – A model "simple nervous system." *Neurosignals, 13*(1–2), 5–19. DOI:10.1159/000076155.

Mackie, G. O., Meech, R. W., & Spencer, A. N. (2012). A new inhibitory pathway in the jellyfish *Polyorchis penicillatus. Canadian Journal of Zoology, 90*(2), 172–181. DOI:10.1139/Z11-124.

Marr, D. C. (1982). *Vision: A computation investigation into the human representational system and processing of visual information*. San Francisco: Freeman.

Maturana, H. R., & Varela, F. J. (1980). Autopoiesis: The organization of the living. In H. R. Maturana & F. J. Varela (Eds.), *Autopoiesis and cognition: The realization of the living* (pp. 73–138). Dordrecht: Reidel.

Maxwell, J. C. (1868). On governors. *Proceedings of the Royal Society of London*, *16*, 270–283.

McCulloch, W. S. (1945). A heterarchy of values determined by the topology of nervous nets. *The Bulletin of Mathematical Biophysics*, *7*(2), 89–93. DOI:10.1007/BF02478457.

McGeer, V. (2015). Mind-making practices: The social infrastructure of self-knowing agency and responsibility. *Philosophical Explorations*, *18*(2), 259–281. DOI:10.1080/13869795.2015.1032331.

Miller, E. K., & Cohen, J. D. (2001). An integrative theory of prefrontal cortex function. *Annual Review of Neuroscience*, *24*(1), 167–202. DOI:10.1146/annurev.neuro.24.1.167.

Milner, A. D., & Goodale, M. G. (1995). *The visual brain in action*. Oxford: Oxford University Press.

Minsky, M. (1986). *The society of mind*. New York: Simon and Schuster.

Mishkin, M., Ungerleider, L. G., & Macko, K. A. (1983). Object vision and spatial vision: Two cortical pathways. *Trends in Neurosciences*, *6*, 414–417.

Moreno, A., & Mossio, M. (2015). *Biological autonomy: A philosophical and theoretical inquiry*. Dordrecht: Springer.

Mundale, J. (2001). Neuroanatomical foundations of cognition: Connecting the neuronal level with the study of higher brain areas. In W. Bechtel, P. Mandik, J. Mundale, & R. S. Stufflebeam (Eds.), *Philosophy and the neurosciences* (pp. 37–54). Oxford: Blackwell.

Nielsen, K. (2010). Representation and dynamics. *Philosophical Psychology*, *23*(6), 759–773. DOI:10.1080/09515089.2010.529045.

O'Keefe, J. A., & Conway, D. H. (1978). Hippocampal place units in the freely moving rat: Why they fire where they fire. *Experimental Brain Research*, *31* (4), 573–590. DOI:10.1007/bf00239813.

O'Keefe, J. A., & Nadel, L. (1978). *The hippocampus as a cognitive map*. Oxford: Oxford University Press.

O'Keefe, J. A., & Recce, M. L. (1993). Phase relationship between hippocampal place units and the EEG theta rhythm. *Hippocampus*, *3*, 317–330.

O'Reilly, R. C., Petrov, A. A., Cohen, J. D., Lebiere, C. J., Herd, S. A., & Kriete, T. (2014). How limited systematicity emerges: A computational cognitive neuroscience approach. In P. Calvo & J. Symons (Eds.), *The architecture of cognition: Rethinking Fodor and Pylyshyn's systematicity challenge* (pp. 191–225). Cambridge, MA: MIT Press.

Pattee, H. H. (1991). Measurement-control heterarchical networks in living systems. *International Journal of General Systems, 18*(3), 213–221.

Penfield, W., & Rasmussen, T. (1950). *The cerebral cortex in man: A clinical study of localization of function*. New York: Macmillan.

Pitt, D. (2020). Mental representation. In E. N. Zalta (Ed.), *The Stanford encyclopedia of philosophy*. Retrieved at https://plato.stanford.edu/archives/win2013/entries/mental-representation/.

Puhl, J. G., & Mesce, K. A. (2008). Dopamine activates the motor pattern for crawling in the medicinal leech. *Journal of Neuroscience, 28*(16), 4192–4200. DOI:10.1523/JNEUROSCI.0136-08.2008.

Quiroga, R. Q., Reddy, L., Kreiman, G., Koch, C., & Fried, I. (2005). Invariant visual representation by single neurons in the human brain. *Nature, 435* (7045), 1102–1107.

Raichle, M. E., MacLeod, A. M., Snyder, A. Z., Powers, W. J., Gusnard, D. A., & Shulman, G. L. (2001). A default mode of brain function. *Proceedings of the National Academy of Sciences of the United States of America, 98*(2), 676–682.

Roseberry, T. K., Lee, A. M., Lalive, A. L., Wilbrecht, L., Bonci, A., & Kreitzer, A. C. (2016). Cell-type-specific control of brainstem locomotor circuits by basal ganglia. *Cell, 164*(3), 526–537. DOI:10.1016/j.cell.2015.12.037.

Rosen, R. (1972). Some relational cell models: The metabolism-repair systems. In R. Rosen (Ed.), *Foundations of mathematical biology* (vol. 2, pp. 217–253). New York: Academic Press.

Rupert, R. D. (2011). Embodiment, consciousness, and the massively representational mind. *Philosophical Topics, 39*(1), 99–120.

Satterlie, R. A. (2018). Jellyfish locomotion. In *Oxford research encyclopedia, neuroscience*. New York: Oxford University Press. DOI:10.1093/acrefore/9780190264086.013.147.

Schwitzgebel, E. (2019). Introspection. In E. N. Zalta (Ed.), *The Stanford encyclopedia of philosophy*. Retrieved from https://plato.stanford.edu/archives/win2019/entries/introspection/.

Searle, J. R. (1980). Minds, brains, and programs. *Behavioral and Brain Sciences, 3*, 417–424.

Sejnowski, T. J. (2018). *The deep learning revolution*. Cambridge, MA: MIT Press.

Sellars, W. (1956). Empiricism and the philosophy of mind. In H. Feigl & M. Scriven (Eds.), *The foundations of science and the concepts of psychology and psychoanalysis (Minnesota studies in the philosophy of science*, vol. 1, pp. 253–329). Minneapolis: University of Minnesota Press.

Shagrir, O. (2010). Marr on computational-level theories. *Philosophy of Science*, *77*(4), 477–500.

Shaw, P. J., Cirelli, C., Greenspan, R. J., & Tononi, G. (2000). Correlates of sleep and waking in *Drosophila melanogaster. Science*, *287*(5459), 1834–1837. DOI:10.1126/science.287.5459.1834.

Shepherd, G. M. (2016). *Foundations of the neuron doctrine* (twenty-fifth anniversary ed.). Oxford: Oxford University Press.

Sohn, J. W. (2015). Network of hypothalamic neurons that control appetite. *BMB Reports*, *48*(4), 229–233. DOI:10.5483/BMBRep.2015.48.4.272.

Sporns, O. (2010). *Networks of the brain*. Cambridge, MA: MIT Press.

Sporns, O. (2012). *Discovering the human connectome*. Cambridge, MA: MIT Press.

Sterling, P., & Laughlin, S. (2015). *Principles of neural design*. Cambridge, MA: MIT Press.

Takahashi, J. S. (2017). Transcriptional architecture of the mammalian circadian clock. *Nature Reviews Genetics*, *18*(3), 164–179. DOI:10.1038/nrg.2016.150.

Tolman, E. C. (1948). Cognitive maps in rats and men. *Psychological Review*, *55*, 189–208.

Tosches, M. A., & Arendt, D. (2013). The bilaterian forebrain: An evolutionary chimaera. *Current Opinion in Neurobiology*, *23*(6), 1080–1089. DOI:10.1016/j.conb.2013.09.005.

Tulving, E. (1983). *Elements of episodic memory*. New York: Oxford University Press.

Valenstein, E. S. (2005). *The war of the soups and the sparks: The discovery of neurotransmitters and the dispute over how nerves communicate*. New York: Columbia University Press.

van den Heuvel, M. P., & Sporns, O. (2011). Rich-club organization of the human connectome. *The Journal of Neuroscience*, *31*(44), 15775–15786. DOI:10.1523/jneurosci.3539-11.2011.

van Essen, D. C., & Gallant, J. L. (1994). Neural mechanisms of form and motion processing in the primate visual system. *Neuron*, *13*(1), 1–10.

van Gelder, T. (1998). The dynamical hypothesis in cognitive science. *Behavioral and Brain Sciences*, *21*, 615–628.

Voigt, J. P., & Fink, H. (2015). Serotonin controlling feeding and satiety. *Behavioural Brain Research*, *277*, 14–31. DOI:10.1016/j.bbr.2014.08.065.

Watts, D., & Strogratz, S. (1998). Collective dynamics of small worlds. *Nature*, *393*, 440–442.

Weber, M. (2005). *Philosophy of experimental biology*. Cambridge: Cambridge University Press.

Weber, M. (2008). Causes without mechanisms: Experimental regularities, physical laws, and neuroscientific explanation. *Philosophy of Science, 75* (5), 995–1007. DOI:10.1086/594541.

Welsh, D. K., Takahashi, J. S., & Kay, S. A. (2010). Suprachiasmatic nucleus: Cell autonomy and network properties. *Annual Review of Physiology, 72*(1), 551–577. DOI:10.1146/annurev-physiol-021909-135919.

White, J. G., Southgate, E., Thomson, J. N., & Brenner, S. (1986). The structure of the nervous system of the nematode *Caenorhabditis elegans*. *Philosophical Transactions of the Royal Society B: Biological Sciences, 314*(1165), 1–340. DOI:10.1098/rstb.1986.0056.

Winfree, A. T. (1987). *The timing of biological clocks*. New York: W. H. Freeman.

Winning, J. (2020). Internal perspectivalism: The solution to generality problems about proper function and natural norms. *Biology & Philosophy, 35*(3), 33. DOI:10.1007/s10539-020-09749-z.

Winning, J., & Bechtel, W. (2018). Rethinking causality in neural mechanisms: Constraints and control. *Minds and Machines, 28*(2), 287–310.

Woodward, J. (2019). Scientific explanation. In E. N. Zalta (Ed.), *The Stanford encyclopedia of philosophy*. Retrieved from https://plato.stanford.edu/archives/win2019/entries/scientific-explanation/.

Yamins, D. L. K., & DiCarlo, J. J. (2016). Using goal-driven deep learning models to understand sensory cortex. *Nature Neuroscience, 19*(3), 356–365. DOI:10.1038/nn.4244.

Zeki, S. M. (1971). Cortical projections from two prestriate areas in the monkey. *Brain Research, 34*, 19–35.

Acknowledgment

The authors thank the series editor, Keith Frankish, and two reviewers, Michael Anderson and Bryce Huebner, for their extremely helpful comments on earlier drafts of this Element. Ta-Lun Huang thanks Academia Sinica, Taiwan, for supporting his residency as a postdoctoral fellow at the University of California, San Diego during the preparation of this Element.

Cambridge Elements

Philosophy of Mind

Keith Frankish
The University of Sheffield
Keith Frankish is a philosopher specializing in philosophy of mind, philosophy of psychology, and philosophy of cognitive science. He is the author of *Mind and Supermind* (Cambridge University Press, 2004) and *Consciousness* (2005), and has also edited or coedited several collections of essays, including *The Cambridge Handbook of Cognitive Science* (Cambridge University Press, 2012), *The Cambridge Handbook of Artificial Intelligence* (Cambridge University Press, 2014) (both with William Ramsey), and *Illusionism as a Theory of Consciousness* (2017).

About the Series

This series provides concise, authoritative introductions to contemporary work in philosophy of mind, written by leading researchers and including both established and emerging topics. It provides an entry point to the primary literature and will be the standard resource for researchers, students, and anyone wanting a firm grounding in this fascinating field.

Cambridge Elements

Philosophy of Mind

Elements in the Series

Mindreading and Social Cognition
Jane Suilin Lavelle

Free Will
Derk Pereboom

Philosophy of Neuroscience
William Bechtel and Linus Ta-Lun Huang

A full series listing is available at: www.cambridge.org/EPMI